797,885 Books

are available to read at

Forgotten Books

www.ForgottenBooks.com

Forgotten Books' App
Available for mobile, tablet & eReader

ISBN 978-1-332-11388-0
PIBN 10286455

This book is a reproduction of an important historical work. Forgotten Books uses state-of-the-art technology to digitally reconstruct the work, preserving the original format whilst repairing imperfections present in the aged copy. In rare cases, an imperfection in the original, such as a blemish or missing page, may be replicated in our edition. We do, however, repair the vast majority of imperfections successfully; any imperfections that remain are intentionally left to preserve the state of such historical works.

Forgotten Books is a registered trademark of FB &c Ltd.
Copyright © 2015 FB &c Ltd.
FB &c Ltd, Dalton House, 60 Windsor Avenue, London, SW19 2RR.
Company number 08720141. Registered in England and Wales.

For support please visit www.forgottenbooks.com

1 MONTH OF FREE READING

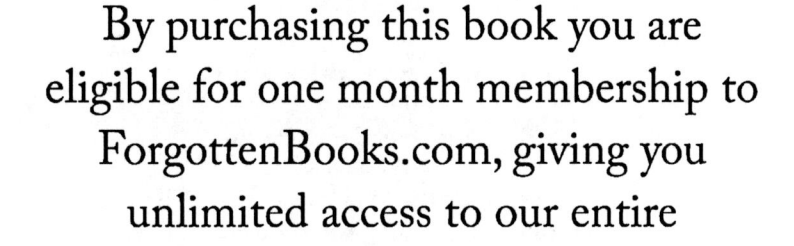

at
www.ForgottenBooks.com

By purchasing this book you are eligible for one month membership to ForgottenBooks.com, giving you unlimited access to our entire collection of over 700,000 titles via our web site and mobile apps.

To claim your free month visit:
www.forgottenbooks.com/free286455

* Offer is valid for 45 days from date of purchase. Terms and conditions apply.

Similar Books Are Available from
www.forgottenbooks.com

The Evolution of Physics
by Albert Einstein

The Quantum Theory
by Fritz Reiche

One Thousand Problems in Physics
by William H. Snyder

Space, Time and Gravitation
An Outline of the General Relativity Theory, by A. Stanley Eddington

Household Physics
by Alfred M. Butler

Theory of Heat
by J. Clerk Maxwell

The Science of Mechanics
A Critical and Historical Account of Its Development, by Ernst Mach

Physics Tells Why
An Explanation of Some Common Physical Phenomena, by Overton Luhr

History and Root of the Principle of the Conservation of Energy
by Ernst Mach

The Physics of the Secret Doctrine
by William Kingsland

A Handbook of Physics
by William Herbert White

A Treatise on Elementary Dynamics
by Sidney L. Loney

The Theory of Optical Instruments
by E. T. Whittaker

Universe, Earth, and Atom
The Story of Physics, by Alan E. Nourse

Scientific Writings of Joseph Henry, Vol. 1
by Joseph Henry

Physics of the Ether
by S. Tolver Preston

A Treatise on Gyrostatics and Rotational Motion
Theory and Applications, by Andrew Gray

The Ideas of Einstein's Theory
The Theory of Relativity in Simple Language, by J. H. Thirring

The Theory of Electricity
by George Henry Livens

Laboratory Exercises in Elementary Physics
A Manual for Students in Academies and High Schools, by Franklin H. Ayres

OPTOMETRY LIB.

H. A. STOC
OPTOME
BERKELEY,

CLINICS IN OPTOMETRY

A compilation of Eye Clinics covering fully all Errors of Refraction and Anomalies of Muscles, with Methods of Examination, Tests and Corrections, as used in actual practice

A Text-Book of the Practice of Optometry

BY C. H. BROWN, M.D.

Graduate University of Pennsylvania; Professor of Principles and Practice of Optometry; Formerly Physician to the Philadelphia Hospital; Author of The Optician's Manual, Vols. I and II.

WITH ILLUSTRATIONS

PUBLISHED BY
THE KEYSTONE PUBLISHING CO.
809-811-813 NORTH 19TH STREET, PHILADELPHIA, U.S.A.
1907

All rights reserved

OPTOMETRY LIB.
repl. M201066

COPYRIGHT, 1907, BY
THE KEYSTONE PUBLISHING CO.

PREFACE

This book has been written not so much for the optical student as for the practitioner of optometry, and it is therefore really a manual of practice. No attempt has been made to commence with elementary matters and systematically pass on to more advanced subjects, but the cases have been taken as they came and as they are likely to be met with in every-day practice. At the same time the explanation of the principles involved in each case has been made as simple as possible, and the method of management made clear, so that even a beginner in the work will have no difficulty in understanding both the theory and the practice in each case.

As these clinics cover the whole range of optometrical practice, the value of this book as a work of reference in difficult cases becomes evident, as the proper method of management can be found in a moment. It is hoped the spirit of this book may inspire the reader to more painstaking and accurate work, and kindle in him an ambition for the advancement of optometry.

CONTENTS

CLINIC NO. 1 — Page
A Case of Simple Hypermetropic Astigmatism 9

CLINIC NO. 2
A Case of Mixed Heterophoria 14

CLINIC NO. 3
Hypermetropic Astigmatism Simulating Myopic 20

CLINIC NO. 4
A Case in Practice with Hysterical Complication—Is Hysteria the Cause or the Result? 25

CLINIC NO. 5
A Typical Case of Hypermetropia 30

CLINIC NO. 6
A Case of Toxic Amblyopia from Alcohol and Tobacco   35

CLINIC NO. 7
A Case of Refractive Error Diagnosed as Cataract . . 41

CLINIC NO. 8
Compound Hypermetropic Astigmatism and Presbyopia Showing when Cylinders May and Should be Omitted from Reading Glasses 46

CLINIC NO. 9
Presbyopia 52

CLINIC NO. 10
Myopia 58

CLINIC NO. 11
Monocular Vision 65

CLINIC NO. 12
Mixed Astigmatism 71

CLINIC NO. 13
Anisometropia 78

CLINIC NO. 14
Convergent Strabismus 85

CLINIC NO. 15
Divergent Strabismus 91

CLINIC NO. 16
Headache in Connection with Myopia and Exophoria   97

CLINIC NO. 17	Page
FACULTATIVE HYPERMETROPIA	103
CLINIC NO. 18	
HYPERPHORIA	109
CLINIC NO. 19	
A CASE OF ASTIGMATISM, ILLUSTRATING THE VALUE OF THE OPHTHALMOMETER	115
CLINIC NO. 20	
A CASE OF HYPERMETROPIA, ILLUSTRATING THE FOGGING METHOD	122
CLINIC NO. 21	
ASTIGMATISM WITH THE RULE	120
CLINIC NO. 22	
ASTIGMATISM AGAINST THE RULE	134
CLINIC NO. 23	
LENTICULAR ASTIGMATISM	141
CLINIC NO. 24	
A CASE OF HIGH MYOPIA	148
CLINIC NO. 25	
KERATOCONUS OR CONICAL CORNEA	155
CLINIC NO. 26	
THE VALUE OF RETINOSCOPY	163
CLINIC NO. 27	
A CASE OF PIGMENTARY RETINITIS, ILLUSTATING THE VALUE OF OPHTHALMOSCOPY	170
CLINIC NO. 28	
ALBUMINURIC RETINITIS, OR THE RETINITIS OF BRIGHT'S DISEASE	177
CLINIC NO. 29	
ACCOMMODATIVE ESOPHORIA	184
CLINIC NO. 30	
SPASM OF ACCOMMODATION	191
CLINIC NO. 31	
EXOPHORIA	199
CLINIC NO. 32	
TRANSPOSITION OF LENSES AS ILLUSTRATED BY A CASE OF MIXED ASTIGMATISM	207
CLINIC NO. 33	
ADJUSTMENT OF SPECTACLES	215
CLINIC NO. 34	
FITTING OF EYEGLASSES	223
CLINIC NO. 35	
SPECTACLES AND EYEGLASSES	231
CLINIC NO. 36	
INSPECTION OF SPECTACLES AND EYEGLASSES AND NEUTRALIZATION OF LENSES	240

CLINICS IN OPTOMETRY

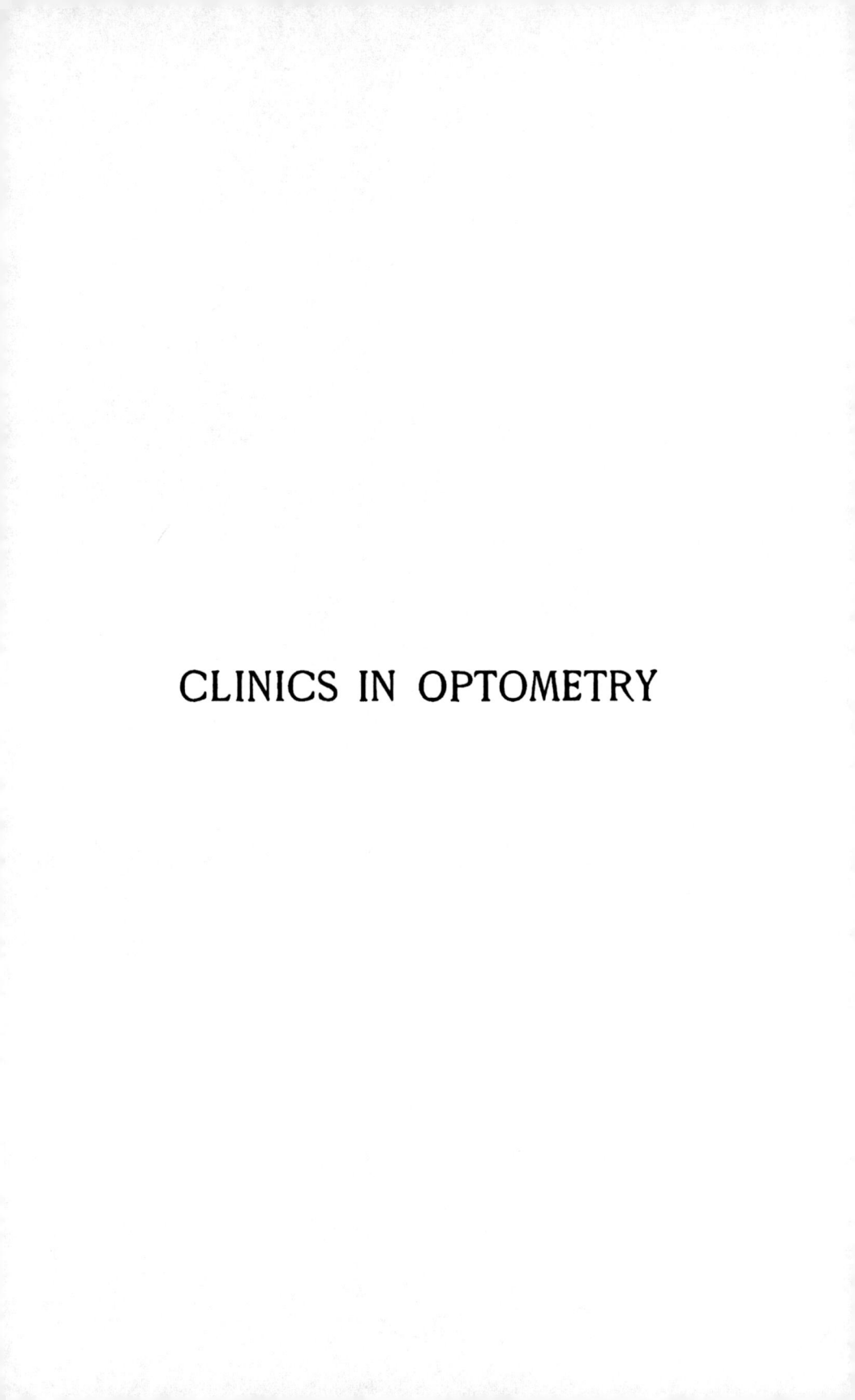

A Case of Simple Hypermetropic Astigmatism

[Clinic No. 1]

J. S. E., a young man, twenty-five years of age, was sent by his family physician. The symptoms of which he complains are as follows: Trouble in reading, especially at night, words run together, pains in head and drowsiness.

The question was asked him, as should be done in every case that applies for relief, whether or not glasses had previously been worn. He replied that he was given glasses some four years ago by an optician, but they were not comfortable, and he never could wear them. On examination by the lens measure they proved to be O. D., — .25 D. S. ◯ — .25 D. cyl. and O. S., — .25 S.

We ask the patient to be seated and direct his attention to the test card hanging on the wall directly facing him, and at a distance of twenty feet. "How far down the card can you read?" is the first question, and in reply he names every letter on the No. 20 line. We make the proper note in our record book that vision $= \frac{20}{20}$, which means that with both eyes open while seated at a distance of twenty feet, he is able to read the line that is marked 20, which indicates that the acuteness of vision is up to the normal standard.

Such being the case we involuntarily make a mental criticism of the concave glasses which had been prescribed by the former optician, and we can readily understand why they never were comfortable. We want to treat everyone fairly and we want to be treated fairly ourselves, and therefore we do not think it wise to condemn our competitors. One can scarcely improve his own reputation by making damaging statements about the work of others engaged in the same line of business as ourselves, even though such statements are entirely truthful. It is far better to make no comment, to say nothing.

But we will leave our patient for a moment, just long enough to preach a little sermon, using this optician's error as our text. Let us take a few moments to consider this matter.

When we find the visual acuteness to be $\frac{20}{20}$, what information is to be gained from this fact? Presumably, the refraction is emmetropic, but *it may be* hypemetropic, or slightly astigmatic, that is, hypermetropic astigmatism. *It cannot be myopic.* Normal vision positively precludes myopia or myopic astigmatism. Therefore, in a case like this, concave lenses must not be tried; they must not even be thought of.

How did the other optician commit such a grievous error? for the prescribing of concave lenses in the presence of emmetropia or hypermetropia cannot be designated in any other way. Perhaps he took the weak concave lenses from his trial case and held them before his patient's eyes, and of course they were accepted. Almost any pair of eyes can see through weak concave lenses at a distance. But the point we want to impress is that concave lenses must not be tried when the vision is normal, and if every optometrist will burn this rule into his brain, it will act as a danger signal to prevent him from straying into paths of error, and also as a guide post to keep him in the straight and narrow path of optometric rectitude.

TESTING WITH TRIAL LENSES

We turn now to our patient and commence the test with trial lenses. We place the opaque disk over the left eye, the right eye being uncovered and ready for the examination. We have made the statement above that concave lenses should not be tried when the visual acuteness is normal. Perhaps we had better make a broader rule as a guide, and that is in every case always to try convex lenses first. In accordance therewith we place a $+.50$ D. sphere in front of the right eye. The patient does not reject it absolutely, neither does he seem inclined to accept it; he is in doubt whether there is an improvement or not. Then we take this sphere out and quickly replace it with a $+.50$ D. cylinder, with its axis at $90°$, so that the patient can mentally contrast the effect of the two lenses, and he at once indicates his preference for the latter as being the best. We repeat the trial of these two lenses, placing first one and then the other before the eye, and the patient unhesitatingly prefers the cylinder, which we then slowly rotate from the vertical meridian first to the right and then to the left. The patient is all the time looking at the No. 20 line of letters on the test card, and he says, as the axis is rotated from the 90th meridian, that the letters become less distinct, and after several trials we conclude that the proper place for the axis of the cylinder is at $90°$. We then place alternately in front of this lens a $+.25$ cyl. axis $90°$, and a $+.25$ S., both of which are rejected. In order to verify the rejection of the cylinder, we remove the + .50 cyl. and place in its stead a + .75 cyl., with its axis in the same position, when the patient says the former is the better. We now feel pretty sure that the + .50 D. cyl. axis 90° is the proper correcting lens. This represents a case of simple hypermetropic astigmatism with the rule.

Perhaps some one may ask why we did not try the card of radiating lines in order to detect the presence of astigmatism in this case. We reply that we do not consider this a reliable test in a case of slight hypermetropic astigmatism, because the ciliary muscle is so quickly brought into action, and by the irregular contraction of which it is said to be capable the defective meridian is neutralized. Theoretically, when astigmatism is present, on account of the difference in refraction of the two chief meridians, the lines running in one direction seem blacker and more distinct than those in the direction at right angles, and therefore when a variation is noticed in the distinctness of the radiating lines, we say astigmatism is present. This difference depends upon the fact that the light entering the emmetropic meridian is focused upon the retina and is distinct, while the rays entering the ametropic meridian are focused in front of or back of the retina, and hence are indistinct. But if the accommodation is brought into action and the hypermetropic meridian thus made emmetropic, both meridians will focus upon the retina, and consequently no difference will then be discernible between the radiating lines.

Another reason why we place but little dependence upon this test for astigmatism, is that in the lower degrees of the defect, where the differences between the lines can be but slight, the patient may be unable to detect them. So much depends upon the intelligence of the patient, and there are so many patients whose powers of observation are but little developed, that the value of this test in the majority of cases is doubtful.

Instead of the card of radiating lines we use the card of test letters, and practically determine the existence of astigmatism by the acceptance of cylindrical lenses. If the patient is in doubt whether cylinders are better than spheres, and if the rotation of the cylinder makes vision neither better or worse, astigmatism is probably absent. But if the cylinder is quickly accepted, and if its rotation produces a decided impairment of vision. we may safely conclude that the case is one of astigmatism.

We now try the left eye. The + .50 sphere is promptly rejected, as is also the + .50 cylinder, with its axis placed at 90° Before removing the latter, however, we try the effect of rotation. As the axis is turned to the right vision is made noticeably worse, but as it is turned to the left an improvement is perceptible, although patient says that vision is clearer without either of these lenses. We now have a clue, however; the fact that vision is improved by rotation of the cylinder in one direction, and impaired by turning the other way, indicates astigmatism. We now try a + .25 cylinder, and as we expected it is at once accepted, and after a few trials we find that the proper position for the axis is at 70°.

We then try the range of accommodation and find the patient can read the smallest print from 4½ to 30 inches. By calling to mind what the normal near point should be at this age, we know that 4½ inches is about right for this age, and that therefore the amplitude of accommodation is not impaired.

MUSCLE TEST BY MADDOX ROD

We pass on then to an examination of the muscular equilibrium, and for this purpose we prefer the Maddox rod, which is placed over the left eye in a horizontal position. We say to the patient that he will see a reddish vertical streak as he looks at the light across the room, and we ask him on which side of the light the streak appears to be. He replies that it is on the left side and about two inches away from it. We mentally analyze this condition: the streak on the same side as the eye, over which the rod is placed, means a homonymous diplopia, is due to esophoria, and is correctible by a prism base out. We therefore place a prism of 1°, with base out, over the right eye, and now the patient says that the streak runs up and down through the light, and we make a note in our record book that there is an esophoria of 1°. This is a not uncommon condition to find, and it is too slight to call for correction.

We then turn the rod around to the vertical position, which causes the streak to be seen horizontally, and we ask the patient what position it assumes with reference to the light, and he replies that it passes directly through it. This indicates that the superior and inferior muscles are equally balanced and that there is no hyperphoria.

Inasmuch as the eyes are apt to vary slightly from day to day, it is well in almost any error of refraction, but especially in astigmatism, to repeat the examination. We therefore ask our patient to return to-morrow, explaining to him the care that should be taken in prescribing glasses. When he returns we make another examination along the same lines as before, and obtain the same result. We therefore feel that we are justified in prescribing the lenses we found as follows :

$$O. D., + .50 \text{ D. cyl. axis } 90°.$$
$$O. S., + .25 \text{ D. cyl. axis } 70°.$$

Which we advise to be worn constantly.

If the second examination had shown a different result from the first, we should have asked our patient to return a third time.

A Case of Mixed Heterophoria

[CLINIC NO. 2]

Mr. R. P., 29 years of age, bank clerk. The symptoms of which this gentleman complains are indistinctness of vision, especially at night, with occasional pain in eyes and head.

In answer to our inquiry as to whether he has ever worn glasses, patient replies that he was fitted with glasses about eight years ago, but he has never worn them much because they were of no benefit to him. We ask to see them, and on neutralization we find them to be + .50 D. spherical.

We now proceed to determine the acuteness of vision, and we find that each eye separately can read all the letters on the No. 20 line, and hence we record the visual acuity as follows: O. D., $\frac{20}{20}$; O. S., $\frac{20}{20}$. We then ascertain the range of accommodation, and we find that he can read the smallest type as close as 4 inches and as far away as 20 inches. This near point of 4 inches shows an amplitude of accommodation of 10 D. We recall that this corresponds to the conditions usually found at 20 years, while at 30 the near point normally recedes to 5½ inches, which represents an amplitude of accommodation of 7 D. This indicates a vigorous condition of the accommodation in this case when the eyes are used for close vision.

INFERENCE FROM THE SYMPTOMS

The symptoms complained of lead us to suspect hypermetropia or hypermetropic astigmatism. In order to get an idea of the condition of the refraction we make a hasty trial with convex lenses held before the two eyes at one time, and we find that a pair of + .50 are accepted. This indicates a hypermetropic condition of the refraction, and proves that there is no tonic spasm of accommodation in spite of the fact that the ciliary muscle is of excessive strength, as shown by an amplitude of 10 D. at 29 years of age.

We will now test each eye separately, and we find that the vision of the right eye is even better than $\frac{20}{20}$, and that some of the letters on the No. 15 line are legible. We try a + .50 D. sphere and it is promptly rejected. We then try a + .25 D. sphere which is accepted. We remove this sphere and replace it with a + .25

D. cyl., which we find is also accepted when the axis is placed at 180°. In comparing these two lenses and quickly changing from one to the other, patient is unable to choose between them. We therefore make a note in our record book that the refraction of right eye is represented by + .25 D. S. or + .25 D. cyl. axis 180°.

On examining the left eye we find the vision is not quite $\frac{20}{20}$ full. A + .50 D. sphere is rejected, but a + .50 D. cyl. axis 90° is accepted. We rotate the cylinder towards the left and patient says letters on the test card are made worse. As we rotate back towards 90° they begin to improve, and keep on until patient again says the letters begin to blur, which does not occur until the axis has passed 20° or 25° to the right from the vertical. After a few trials we find that 110° is the best position for the axis of the lens.

We next make an examination with the ophthalmometer. The right eye shows an overlapping of half a step in the vertical meridian. Now, it will be remembered that in the normal cornea there is an excess of curvature and of refraction in the verical meridian of about this amount, and therefore in astigmatism with the rule the same allowance must be made; consequently this eye, as far as the curvatures of the cornea indicate, is devoid of astigmatism.

The left eye shows an overlapping of one and a quarter steps (1.25 D.) in the vertical meridian: after making the usual deduction of .50 D. or .75 D. the ophthalmometer indicates an astigmatism of at least .50 D. with the rule.

We feel safe now in deciding that there is practically no astigmatism in the right eye, and that its probable refraction is represented by a + .25 D. sphere.

The left eye is undoubtedly astigmatic to the extent of + .50 D., the excess of curvature being at or near the vertical meridian. The ophthalmometer shows the axis of the cylinder at 90°, while the test with the cylinders calls for the axis at 110°. We would feel better satisfied if both tests agreed as to the location of the axis; but this is a discrepancy that often occurs, and in such a case we must be guided by the answers of the patient who, if of average intelligence, after a few trials will be able to decide in which position the cylinder affords the best vision. We therefore repeat the test with the cylinder by rotating it from one position to another, when we find that Mr. P. adheres to his statement that the axis at 110° affords the best vision.

What is the result of our examination of the refraction? O. D., + .25 D. S. ; O. S., + .50 D. cyl. axis 110°

USE OF MADDOX MULTIPLE ROD

The next step in the examination is the use of the Maddox multiple rod to determine the condition of the muscular equilibrium. We place it in a horizontal position over the left eye and ask the patient if he sees a red streak of light running vertically, and if so what position does it assume with regard to the light seen by the uncovered eye. He replies that he sees it and that it is about one inch to the left of the light. This indicates esophoria of low degree. We try a prism of 1°, with base out, and find this brings the red streak of light directly through the flame.

We turn the Maddox rod around to the vertical position, when the patient sees the red streak running horizontally, and we ask him what position it assumes with regard to the light, whether above or below or through it, and he replies that the red streak runs directly through the center of the light. This indicates a proper balance of the vertical muscles, and we conclude there is no hyperphoria. The dot and line test shows 2° of exophoria in accommodation.

This concludes the examination for the day, and in accordance with our usual custom we ask the patient to return for a further examination, telling him we think it advisable to make at least three examinations before prescribing glasses.

On the second visit we get exactly the same result as on the first day.

But on the third day we uncover a condition of latent heterophoria. We measure the power of convergence and of divergence as a part of our routine method, and find the former is equal to 28°, which is normal, while the latter is equal only to 3°, which is considerably below normal.

For the benefit of our less experienced readers we will say that the power of convergence is measured by the strongest pair of prism, bases OUT, which the eyes are able to overcome and maintain the light single at a distance of 20 feet. The normal power is from 20° to 30°, but which can be trained up to 60°.

The power of divergence is measured by the strongest prisms, bases IN, which the eyes are able to overcome and maintain the light single. The normal standard is from 6° to 8°.

These tests show in this case a weakness of the divergent muscles, the eyes being given over to the stronger convergent muscles, which causes a condition of esophoria.

After these duction tests, when each muscle is called upon to exert itself to the utmost, any existing spasm of a single muscle is likely to be broken up, and the true condition of the several muscles can the better be determined. We now try the Maddox rod on this case, and instead of the 1° of esophoria which we found on the first two examinations, there is now an esophoria of 10°, and a right hyperphoria of 1° has become manifest.

Instead of prescribing glasses on the third examination, as we expected, we tell our patient that there have been some new developments in the case which make it more complicated, and that several more examinations will be required in order to clear it up.

These tests resulted as follows:

Oct. 31, 1903. Esophoria 10°. Right hyperphoria 1°.
Nov. 2, " " 4°. " " 1°.
" 3, " " 15°. " " 2°.

The first test showed an exophoria of 2° at the reading point, but the later tests of the muscular balance at the near point showed the following results:

Oct. 31, 1903. Esophoria 20°
Nov. 2, " " 5°.
" 3, " " 13°.

It is a well-known fact that the muscular balance or imbalance is a variable quantity, and no two days show exactly the same results. It is also a fact that there may be a spasm of one of the extra-ocular muscles, which masks the true condition, as in this case the test of the muscles at the reading point at first showed an exophoria, when really the condition is one of esophoria.

PROCEDURE IN CORRECTION

The later developments in this case lead us to suspect that the symptoms are perhaps as much due to the heterophoria as to the error of refraction. Ordinarily we think it proper to correct the refractive error first. Oftentimes such glasses will afford perfect relief to all the symptoms, and by relieving the strain on the accommodation the heterophoria gradually passes over into a condition of orthophoria. If, however, the symptoms continue unabated in spite of the spheres or cylinders that have been worn, then we advise the combination of prisms for the correction of any existing muscular insufficiency.

But in this case there is such positive evidence of spasm of the external recti, and such a high degree of esophoria after this spasm had been removed, that we think it best to give him a prismatic correction at once.

The amount of esophoria varied considerably from day to day, and therefore we hardly feel justified in prescribing a prism stronger than 3° for its correction. The hyperphoria was constant after it first made its appearance; on three days it was 1°, and on one occasion it reached 2°. We will therefore prescribe a 1° prism for the correction of the vertical deviation.

The right eye is patient's best eye, and as the amount of hypermetropia is so slight we will not correct it. We will place the 1° over this eye, a plane prism without any spherical or cylindrical curvature.

The astigmatism in the left eye cannot be ignored, and therefore we will prescribe the indicated cylinder combined with the 3° prism.

Our prescription reads:

O. D., Prism 1°, base down.
O. S., + .50 D. cyl. axis 110° ◯ prism 3°, base out.

Perhaps some one may wonder why we place the weaker prism over the right eye. We have a very good reason for this; we want to interfere as little as possible with the good eye, and this we do by placing the 1° prism over it. The 3° prism causes a more marked displacement of objects, which would be much more noticeable, and hence more annoying, if placed over the good eye than if placed in front of the other eye. If a single prism is prescribed it should always be placed over the eye with the least acuteness of vision. If the optometrist will keep this in mind he may save himself annoyance and his patient discomfort, and he may be able to get more satisfaction out of the use of prisms.

These glasses were prescribed, and they seem to have filled the bill completely. It is now six months and more since Mr. P. commenced to wear them, during which time we have seen him on several occasions on account of breaking of the left lens, and he always speaks in the highest terms of the comfort and relief afforded by the glasses.

Perhaps it might be well to say that prismatic lenses should not be set in frameless mountings on account of the liability to breakage at the apex end of the prism, which is very thin. We advised this patient to have frames, but he preferred frameless mountings with the result of two broken lenses since he has been wearing them.

Hypermetropic Astigmatism Simulating Myopic

[CLINIC No. 3]

Mrs. I. A., aged twenty-two, complains of headache, soreness over eyes and twitching of lids.

We find the acuteness of vision is $\frac{20}{40}$ in each eye.

In a case like this where headache is the chief symptom, and where the acuteness of vision falls somewhat below normal, we suspect astigmatism. After trial of lenses, in which all convexes are rejected, we find that — 1 D. cyl. axis 180°, is accepted for each eye separately, and raises the visual acuity to $\frac{20}{30}$. We therefore apparently have a case of simple myopic astigmatism; but we feel entirely satisfied with the result of our first examination, not enough so to justify us in prescribing glasses, and therefore we will pursue our investigation further.

We give her the small reading card in order to measure the amplitude of accommodation, and find that she can read the smallest print at the top of the card as close as 7 inches and as far away as 20 inches. The power of accommodation is therefore 5.50 D. We recall the normal near point at this age should be 4½ inches, representing 9 D. of accommodation. We are impressed at once with the fact that the near point has receded and that the amplitude of accommodation has lessened, which indicates that the refraction of the eyes is hypermetropic, instead of myopic, as the cylinders chosen would lead us to believe.

Let us stop long enough to remind the reader that in hypermetropia there is a deficiency of refractive power, which must be made up by the accommodation; this reduces the amount of available accommodation as revealed by the receded near point. In manuals of refraction will be found several tables showing the position of the near point and the amount of accommodation at the different ages. It is well for the optometrist to have these tables committed to memory, so that knowing the age of his patient he can quickly and mentally note if there is any departure from the normal standard and whether in the direction of deficiency or excess of accommodation.

Contrariwise in myopia there is an excess of refractive power, thus increasing the amplitude of accommodation, which surplus is revealed by a near point closer than the normal standard for that particular age.

Therefore the receded near point in this case and the diminished amplitude of accommodation, would tend to show that the concave cylinders accepted were incorrect and would rather indicate a hypermetropic condition of refraction.

DETERMINING MUSCULAR EQUILIBRIUM

We now pass on to determine the muscular equilibrium, for which purpose we place a Maddox rod over the left eye in a horizontal position, while the patient's attention is directed to the light on the other side of the room. On account of the dissimilarity of the images formed on the two retinæ, the natural desire for binocular fusion is interfered with, and the eyes are given over to the action of the muscles uninfluenced by this instinct, and therefore if there is any imbalance it at once becomes manifest.

The image formed in the right eye is that of the light of its natural appearance, while the image formed in the left eye is a vertical streak of light of a reddish tint.

We ask patient if she sees a red streak of light, and she replies that she does not. This is by no means unusual, and the young optometrist need not be worried because the patient fails at first to see the streak and therefore he is led to fear that he may not be able to make use of this test, which is really the most reliable at our command in detecting the presence of heterophoria. The image of the natural light formed in the right eye is bright and clear and engages the attention of the brain to the temporary exclusion of the less distinct image formed in the eye that is covered by the Maddox rod, because the latter makes a much less marked impression upon the visual centers.

We cover the right eye for a second and then the streak at once becomes visible, and on removing the cover the streak can usually be retained in the visual field; or perhaps the better way is to rotate the Maddox rod, which at once engages the attention of the left eye and makes the streak noticeable. And whenever the patient loses the streak, as they oftentimes do, one or two rotations of the rod at once brings it into view.

We make use of this little trick with this lady and we ask her again if she sees a vertical streak, and now she replies in the affirmative. We ask her on which side of the light the streak is located, and she replies "to the left." Now, let us consider what we have found: the rod is over left eye, and streak is seen to left side; this is the form of diplopia that is termed homonymous; that is, right object is seen by right eye and left object by left eye, and indicates a condition of esophoria.

USE OF PRISMS

We ask the patient how far the streak appears to be from the light, and she replies "about three or four inches," which amount of displacement indicates at least several degrees of defect. We recall that in esophoria the base of the prism must be placed out, and we hold a prism of 2° in this position before the eye and ask patient what effect it has on the relative positions of the streak and the light. She replies "that the streak is still on the left of the light, but closer to it, perhaps within an inch or two."

We then try a 3° prism in place of this one, and patient tells us that while streak is still on the left, it is now almost in contact with the light. We feel as if this is about the measure of the esophoria, but before entering the record in our case book, we will try a 4° prism base out, and in answer to our inquiry patient says the streak is now at least one inch to the right. This, of course, is an over-correction, and we record the insufficiency as 3°.

Now, it must be remembered that esophoria is usually associated with a hypermetropic condition of refraction. When we find hypermetroia we may safely expect a tendency to inward deviation, and when the muscle tests reveal esophoria, we are almost sure to find that the refraction of the eye is hypermetropic.

Therefore in this case, in spite of the fact that concave cylinders were preferred, the diminished amplitude of accommodation and the esophoria would both throw a shadow of doubt over the accuracy of these lenses.

APPLICATION OF FOGGING SYSTEM

We will now make use of the fogging system in our effort to detect some of the hypermetropia which, we think, is existing in this case in a latent form. We reduce the strength of the convexes by concaves, until now we have a plus value of 1 D. before the eyes, with which, however, vision is very much blurred and indistinct. Instead of reducing still further with concave spheres, we will try concave cylinders; placing them in the trial frame with the position of the axis at 180°. We commence with a — .50 cyl. which produces an immediate improvement in vision; we try successively a — .75 cyl. and a — 1 cyl. each improving over the other, and the latter affording a vision of $\frac{20}{30}$ full and $\frac{20}{20}$ partly. The axis of the cylinder is rotated first in one direction and then in the other, but patient prefers the position of the axis at 180° as affording the best vision. This was the result of the test of the right eye, and on the left eye a — .75 cyl. was preferred with axis at 180°. Now let us see what we have ·

    O. D., + 1 D. sph. ◯ — 1 D. cyl. axis 180°
    O. S., + 1 D. sph. ◯ — .75 D. cyl. axis 180°

Transposing these to simpler forms we get

    O. D., + 1 D. cyl. axis 90°
    O. S., + .25 D. sph. ◯ + .75 D. cyl. axis 90°

We feel now that we have worked out the proper correction, and we want to use this case as a text from which to preach a short sermon to optometrists, and particularly to the younger and less experienced men in the practice, that they must be very careful not to mistake hypermetropia for myopia, and that they must be slow in prescribing concave lenses, unless the indications for them are unmistakable.

It is well to remember that in the great majority of cases that apply to you, the refraction is hypermetropic. In law, a man is assumed to be innocent unless he is proven guilty. In optometry we would lay down and establish the broad general principle that the error of refraction in any case that applies to us is assumed to be hypermetropic unless there is positive and undeniable proof to the contrary.

We are not likely to give convex lenses when myopia is actually present; and if we did, no great harm could be caused by them. But we *are* likely to fall into the error of giving concave glasses when the refraction is really hypermetropic (this has been done over and over again by men who should know better), thus imposing a tax on the accommodation and producing a condition of asthenopia, with symptoms of great discomfort.

Our advice in cases of astigmatism, or in any case where there is difficulty in determining the exact condition of refraction, is not to prescribe glasses on a single examination, but to repeat it for a third time. We did so in this case and arrived at practically the same result each time, but we did not think it wise to order the full correction at first, therefore we compromised on the following·

O. U., + .50 D. cyl. axis 90°

These lenses were prescribed in May, 1903, and have given entire satisfaction. Patient has called several times since when lenses were broken, and always speaks in the highest terms of benefits received from the glasses.

A Case in Practice with Hysterical Complication—Is Hysteria the Cause or the Result?

[CLINIC No. 4]

Young lady, sixteen years of age, complains of headache. Applied for relief two years ago, at which time she was wearing + .50 D. cyl. axis 90° over each eye. These glasses were fitted by an oculist, and although she returned to him twice, complaining, he announced them correct.

She states that in addition to the headache, her eyes always ached more or less, and that artificial light was extremely trying to them. These symptoms were aggravated by close use of the eyes.

An intra-ocular examination showed the fundus to be normal. Visual acuteness, O. D. $\frac{6}{9}$, O. S. $\frac{6}{9}$. As subjective test was unsatisfactory, reliance was placed upon the retinoscope, which indicated a hypermetropia of .75 D. With these lenses vision of each eye singly was $\frac{6}{9}$, but binocular vision was $\frac{6}{6}$ clearly.

These glasses were prescribed with instructions to wear constantly, and to report in one month, at which time she returned and said: "Head little better, but still have trouble."

An examination at this time showed, without glasses, O. D. $\frac{6}{9}$??, O. S. $\frac{6}{9}$??, O. U. $\frac{6}{9}$. + 1 D. was indicated by the ophthalmoscope, with which vision was O. D. $\frac{6}{12}$, O. S. $\frac{6}{12}$, O. U. $\frac{6}{9}$.

Ophthalmometer showed an overlapping in vertical meridian of about half step in both eyes, which is only the normal amount of excess of corneal curvature.

Muscle tests by Maddox rod and Maddox double prism showed an exophoria of 4° at distance, and of 8° at reading point.

+ 1 D. was prescibed for constant wear. In two weeks she returned and said: "Headache gone, but light still troubles me."

Muscles were retested with the same result as before, and as a partial correction for the exophoria, a prism of 1°, base in, over each eye was combined with the spherical correction. These glasses afforded instant relief and freed her from the asthenopic symptoms.

PHYSICIAN DIAGNOSED HYSTERIA

In six months she came back with the following statement ·
She had been vaccinated, two weeks after which all her eye
symptoms returned in an aggravated form. Another examination
was made and finding the eyes about as they were at the previous
visit, advised the lady to consult a physician, who pronounced it a
case of hysteria, with various symptoms other than ocular, for which
she was sent to the hospital for treatment.

Her general health was much improved, but she came out
of the hospital still suffering with the head trouble. She consulted the optometrist again, and as another examination
revealed nothing further, she was advised to apply to a different
oculist.

Nothing more was seen of her until January, 1904, when she
affain returned, wearing the glasses last prescribed, complaining of
headache and the glasses seeming smoky. An examination at this
time showed O. D. $\frac{6}{6}$, O. S. $\frac{6}{6}$. Retinoscopy $+ .50$ D., with which
vision was same as without glasses.

The optometrist consulted now was very much surprised to
find the accommodation was lacking, as it required a $+ 2.75$ D.
addition to distance correction to enable patient to read at 14 inches,
while with distance glasses alone could barely read 1.75 D. type at
35 cm., and that for a moment only.

Muscle tests resulted as follows: At 20 feet, 3° exophoria;
at 13 inches, 7° exophoria. Adduction 19°, abduction 12°. The
former prescription, $+ 1$ D. S. $\bigcirc$ 1° prism, base in, was continued
for distance, and extra fronts containing $+ 2.75$ D. given her for
near work.

As her head trouble was not relieved, he concluded that the
symptoms originated from hysteria, and again referred her to a
medical specialist for treatment. Atropine was prescribed, but as
this did not meet his sanction, she was sent to another oculist, who
said she did not need glasses and advised her to leave them off.

After two months she returned with the statement that her
headaches had grown steadily worse. Vision now was only $\frac{6}{12}$;
refraction same as before, but the glasses afforded no improvement
in vision as they previously did.

The optometrist is fully convinced that the whole is due to
hysteria, but that the deterioration of vision revealed in the last

examination is in addition partly attributable to the non-wearing of glasses.

WHAT THIS CASE TEACHES

We will study this case together as it stands before us, and see if anything practical can be learned from it.

We are not quite ready to agree with the diagnosis, but granting that this young lady is hysterical, is hysteria caused by the eye-strain, or to reverse the question, are the ocular symptoms the result of the hysteria? The one proposition may be just as true as the other. The optometrist seems to think the eye symptoms are caused by the hysteria, but he offers no arguments to prove his position. We ask why may not the hysteria (if such exists), be due to the ocular conditions? At any rate, we think the proper plan of treatment to be followed is a more determined effort to correct the refractive and muscular anomalies.

All that was done for the exophoria was the prescription of a pair of 1° prisms. Was this sufficient? We think not.

The first step in the treatment should be the correction of any existing error of refraction. Now it must be remembered that exophoria is usually associated with myopia, in which case a correction of the latter lessens the former. Therefore, if concave lenses lessen exophoria, convex lenses will increase it. Now the lesson to be deduced from this fact, and applied in the case before us, is to give the convex lenses as weak as possible, because the stronger the convex lenses that are worn, the greater will be the exophoria.

This fact brings out another feature of this case, viz.: that the functions of accommodation and convergence do not bear their normal relation to each other. If hypermetropia was present, as the tests show, and if accommodation and convergence bore their customary close relation, the accommodation that would be used to overcome the hypermetropia would call the convergence into action and produce an esophoria. Instead, in this case, we find exophoria, which complicates the case—makes it more difficult of correction, and helps to account for the asthenopic symptoms.

Therefore, in this case, we would not prescribe convex lenses stronger than .50 D.

EXERCISE BY PRISMS

The next step in the treatment is to make an effort to increase the power of convergence, which in this case is relatively weak.

This is to be accomplished by prismatic exercise of the internal recti muscles. The strongest prisms, bases out, which the patient can overcome, and maintain singleness of vision, should be placed in a frame and worn for several minutes, lifting them from the eyes occasionally during this period while gazing at a light across the room. The length of the exercise should not exceed five minutes, and it may be repeated daily. The record of the adduction in this case is 19°, but it can probably in the course of a month or two be trained to 40° or 50°, by a gradual increase in the strength of the exercising prisms.

This exercise should be conducted in the optometrist's office, in connection with which we would give the patient a pair of 5° prisms, bases out, set in a spectacle frame for home use, with instructions to wear from fifteen minutes to a half hour each day while going about the house.

We seldom find it necessary to order prisms for constant wear in exophoria unless of much higher degree than this case; however, as the 1° prisms, bases in, afforded so much relief in general vision, the optometrist was quite justified in prescribing them. But as exophoria is always greater at the reading distance, and as it is always increased by stronger convex lenses (+ 3.75 being now required for close use), it follows that stronger prisms are indicated for the reading glasses, and therefore the extra fronts should contain not only convex spheres but prisms as well; perhaps + 3 D. sph. ◯ prism 2°, base in, to be placed over the + .50 D. sphere, and 1° prism worn for general vision.

After the optometrist has done all this, that is, got the young lady started on this line of optometric treatment, she may be referred to the medical man for supplemental treatment, which would probably consist of general or nerve tonics.

If, under the influence of this combined treatment, the power of convergence is increased, as we have reason to expect, we may reasonably look for relief from the symptoms complained of, in which case we will be able to reduce the reading prisms.

The effect of the exercise with prisms, bases out, which calls the convergence so strongly into action, will also through their close nervous correction, have a stimulating and beneficial action on the ccommodation, which seems to be too weak to perform its function.

EYE MANIFESTATION OF HYSTERIA

Let us for a moment consider what ocular symptoms usually result from hysteria.

There may be twitching and spasmodic closure of the lids with spasm of accommodation ; in our case the accommodation is relaxed, or perhaps paralyzed ; therefore there is nothing in this symptom to indicate hysteria ; in fact it rather tends to contra-indicate it.

The visual disorders that are found in connection with hysteria vary from a slight impairment of vision to total blindness. Hysterical amblyopia comes on suddenly, the ophthalmoscopic appearances being negative. Both eyes may be affected, but more frequently the trouble is monocular. The vision is unexpectedly improved by any kind of lenses—convex, concave or even prismatic—acting in all probability through the imagination. Hysterical amblyopia does not usually last long, and disappears as suddenly as it came.

These symptoms do not correspond very closely with those of the case under consideration, and therefore we conclude that this is not a typical case of hysteria, and that the nervous symptoms are more likely the result of the eye troubles than the cause of them.

This young lady should be advised to disregard and ignore her eye symptoms : not to allow her mind to dwell on them, but to keep it occupied with other things to the exclusion of her eyes. In other words, her eye symptoms will never improve as long as her mind dwells on them ; while they are much more likely to disappear if she can forget them.

A Typical Case of Hypermetropia

[CLINIC NO. 5]

R. S. C., aged nineteen years, a hat finisher by occupation. Complains of headache which comes on about noon every day except Sundays. This symptom is a striking one and indicates at once that the headache is due to eyestrain.

At first on going to work in the morning, the ciliary muscle which had been refreshed and invigorated by the previous night's sleep, is easily able to overcome the refractive error and afford good vision without any apparent effort. But after a while this continued tax on the accommodation, which is so much greater than it should be normally, begins to cause fatigue and by the time the noon hour is reached the eye is sending vigorous complaints to headquarters (the brain), that the burden of work imposed upon it is in excess of its natural capacity. This is simply and solely what the headache means.

The normal eye, as is the case with every other organ in health, performs its function without our consciousness. As soon as the eye complains, in other words, as soon as feelings of discomfort remind us that we have eyes, there is evidence that they are not as they should be, that something is wrong. These warnings must be heeded, and they are by the prudent man; if disregarded, the trouble grows worse and the condition of the eyes may become serious.

DIAGNOSING THE CASE

So far we have this symptom of headache following use of the eyes, and we naturally think of hypermetropia or hypermetropic astigmatism. We ask our patient what letters are legible on the distant test card, and we find that the acuteness of vision in each eye is $\frac{20}{15}$. This is so much better than normal that it practically excludes astigmatism.

In the further determination of this point, we make ophthalmometric examination, and find that the mires overlap one-half step in the vertical meridian in each eye. This corresponds to the normal corneal curvature, and is corroborative of the opinion expressed a moment ago that the error is not one of astigmatism.

This leaves hypermetropia as the probable condition of refraction, and in order to determine its existence we place convex lenses before the eyes as the patient views the distant test card, and we find that $+1$ D. lenses are readily accepted, thus indicating a manifest hypermetropia of this amount.

The hypermetrope (if not too old and if the error is not of too high degree) usually sees well at a distance, in fact he oftentimes enjoys an unusual degree of acuteness of vision. Such a person, in view of his remarkable vision, is apt to laugh at the suggestion of a defect in his eyes, and the thought of glasses is to him little less than ridiculous.

We proceed now to examine our patient's near vision and measure his amplitude of accommodation. We find that he can read the small print as close as 5 inches and as far away as 27 inches; this near point of 5 inches represents an amplitude of accommodation of 8 D. Now how does this compare with the normal standard?

In any treatise on eye refraction will be found a table showing the amount of amplitude of accommodation that should be present at the different stages through life. Every case must be gaged by comparison with this table, and in this way any departure from the standard at any given age can be readily detected.

At twenty years of age the eye should possess 10 D. of amplitude of accommodation; our patient at nineteen possesses only 8 D. of accommodation. There is, therefore, an evident deficiency of 2 D. at least, and a presumption of the existence of hypermetropia of like amount.

It will be remembered that in hypermetropia distinct distant vision is possible only by an effort of accommodation equal in amount to the grade of defect, and hence in close vision there is a deficiency of like amount. Therefore, in this young man's case when we find that the accommodation at the near point is 2 D. less than it should be, we are justified in assuming that this amount of accommodation has been used to overcome hypermetropia and render distant vision clear.

This examination of the amplitude of accommodation corroborates our diagnosis of hypermetropia, and shows it to be even greater than the acceptance of the convex lenses has indicated.

CORROBORATIVE TESTS

Continuing our examination, we look into the condition of the muscular equilibrium. We use the Maddox rod and place it over the left eye in a horizontal position. The image formed in this eye is a vertical streak of light, which our patient tells us is about six inches to the left of the light. This is a condition of homonymous diplopia, due to an over convergence of the eyes, and indicates esophoria. After trying several prisms, we find that one of 4° base out brings the red streak directly throngh the flame. This esophoria which we have discovered in this way, is additional corroborative evidence of the presence of hypermetropia; because the latter, by calling the accommodation into play, also stimulates the convergence; and hence, unless there is some disturbance of the relation that closely binds the functions of accommodation and convergence, esophoria always accompanies hypermetropia.

Our next step in the examination is to try each eye separately with the test lenses, and inasmuch as the lessened amplitude of accommodation indicates a degree of hypermetropia greater than the convex lenses previously accepted, we will make use of the fogging method.

We place $+ 5$ D. in the trial frame in front of the right eye. Patient says he cannot see a single letter on the test card. We hold a $-.50$ D. in front of the $+ 5$ D., and patient says the card looks clearer, but he is still unable to name any letters. We replace the $-.50$ D. by a -1 D. and now patient is able to see the large letter at the top of the test card. We gradually increase the strength of the concave at first by $-.50$ D. and then by $-.25$ D., until finally patient is able to name all the letters on the No. 20 line and most of them on the No. 15 line.

Now, we will see what we have in the trial frame. In the back cell we have the $+ 5$ D. lens which was originally placed there. In the front cell we have $- 2.50$ D. The latter neutralizes half of the former, and leaves a plus value in front of the eyes of 2.50 D., with which patient has normal vision, and therefore we have discovered or unearthed a hypermetropia of this amount.

The convex lens causes the rays of light to enter the eye in a convergent condition. A hypermetropic eye (with suspended accommodation) is adapted for convergent rays, in fact there is no other form of eye in which convergent rays can focus on the retina;

therefore, when a patient can see at a distance through convex lenses, in other words, when convergent rays can be received by an eye, hypermetropia is proven to exist, and theoretically the degree of convergence will correspond to the amount of hypermetropia.

Of course, it is understood that the accommodation is capable of making the entering rays convergent, and thus focusing them on the retina. But this places a burden on the ciliary muscle, for which nature never intended it; thus causing the headache and other asthenopic symptoms; and it is the duty of the optometrist to ascertain if the accommodation is subjected to such unnatural strain, and if so, to take the proper means to relieve it.

We now test the left eye in the same way by the fogging method, and we find that when a — 3 D. is placed in front of the fogging lens of + 5 D., that normal acuteness of vision is secured, thus indicating a hypermetropia of 2 D.

We will finish our examination of this young man's eyes with the retinoscope, which, I am glad to say, gives us the same results as found by the fogging method.

DIAGNOSIS VERIFIED

We now have a complete record of this case, and as all tests agree, we have no difficulty in arriving at a satisfactory diagnosis of the nature of the defect and the amount. This being done, it might seem at first thought that a pair of lenses O. D. + 2.50; O. S. + 2, should be prescribed, to the instant relief and satisfaction of the patient.

But the experience of the writer has been, that it is not well to try to force too strong a convex lens on a young eye that has never been accustomed to wearing glasses. It will blur distant vision and make the eye uncomfortable; of course, if the use of the glasses is persisted in for several months, the impairment of vision and feeling of discomfort will gradually disappear; but very few patients are willing to wear glasses in the face of such disadvantages. When glasses are put on, the patient naturally expects to see better; he thinks that is the purpose for which glasses are worn. If, on the other hand, vision is impaired by the glasses, the impulse of the patient is to take them off, and it is only exceptionally that he can be persuaded to wear them under such circumstances.

For these reasons, in this case, where patient is young and accommodation is vigorous, we will correct only half the error and order glasses as follows ·

O. D., + 1.25 D. ; O. S., + 1 D.

When patient made a report later we were told that these glasses have relieved the headache and afforded entire satisfaction. Whereas, if we had attempted to give a full correction, patient would have returned most likely very much dissatisfied, when we would have been compelled to reduce the strength of the glasses, or patient would have wandered off to some other optometrist, of more tact if not of more skill. Why, then, shouldn't we prescribe the weaker glasses first, and thus maintain our reputation and retain the patient's confidence?

A Case of Toxic Amblyopia from Alcohol and Tobacco

[Clinic No. 6]

The case we now present for your consideration is a man, 51 years of age, a quarryman by occupation. He was sent to us by a most capable refractionist to whom he applied for glasses, and who realized at once that the case was out of the ordinary and that glasses were of doubtful benefit in the present condition of his eyes.

The patient tells us that his eyes feel heavy, that his vision has been failing for some time, and that for the past four weeks he has been unable to read, without glasses or with them.

We ask the man to look at the test card hanging across the room, and we find the acuteness of vision of the right eye is $\frac{20}{100}$, while with the left eye he is unable to decipher even the largest letter at the top of the card.

THE PIN-HOLE TEST

In order to determine whether this impairment of vision is due to ametropia, or whether it depends upon a condition of disease, we will make use of the pin-hole disk. It has the effect of reducing the circles of diffusion formed upon the retina, and thus makes the outlines of objects much clearer.

On account of the smallness of the opening, it diminishes the illumination, but this loss of light is more than compensated for by the greater clearness of form that is attained.

If the vision is susceptible of improvement by this test, glasses will be of benefit. If the vision cannot be improved by the pin-hole, glasses are useless. This test is one that can always be depended upon.

By cutting off all the marginal rays and allowing only a few of the more central rays to fall upon the retina, the eye is rendered independent of its refracting media, in fact, their function is destroyed, and the image is formed solely by the few central rays that pass through the small opening. The image thus being made perfect; if now the vision still continues impaired, it must be because the retina is not capable of receiving, or the optic nerve of transmitting, this image.

Any capable refractionist can demonstrate this for himself. Take any kind or power of lens from your test case, convex or concave, weak or strong, sphere or cylinder, hold it close to your eye and look through it at the test letters; vision will be made more or less indistinct according to the strength and character of the lens, but no matter how greatly the letters are blurred or even if blotted out of sight, when the pin-hole disk is placed over it, the power of the lens is destroyed, and the letters are as clearly seen as with the emmetropic eye.

We will now see what the results of this test are: vision of right eye is little if any improved. Vision of left eye is raised to $\frac{20}{30}$ with some difficulty. This proves that the impairment of vision in R. E. is not susceptible of any improvement by lenses, while L. E. is only partially so. This is a very great diminution in vision and would also indicate his inability to read fine print. We give him the test card in his hand, and as we expected, he says he is unable to read any of the paragraphs. As a matter of routine, we try successively $+1, +2, +3, +4$ and $+5$, but with none of them is he able to read anything except the larger sized print at the bottom of the card.

QUESTIONING THE PATIENT

We will take this man into our dark room and make an ophthalmoscopic examination. The reflex is clear and bright proving that there are no opacities in any of the refracting media. The optic disk looks a little paler than normal with perhaps a bluish or greenish tinge, otherwise no evidence of the presence of disease.

We will question this man as to his personal habits. "Do you drink or smoke?" "Yes, sir." "Now tell us the truth, how much do you drink? Tell these gentlemen just what your habits are in this direction." "Every evening on my way home from work I stop in the saloon and take a drink of whisky and wash it down with a mug of beer. I spend Saturday evening there when I take from three to four to half-a-dozen drinks."

"As this is a matter of importance to your welfare, you must not conceal anything, and I am glad you have given us such a frank answer. Now how about tobacco?"

"I smoke both cigars and a pipe, and am seldom without one or the other in my mouth."

The very candid statement of this patient clears up the case, and I think we can unhesitatingly diagnose this condition as one of toxic amblyopia due to the excessive use of alcohol and tobacco; and as you are likely to meet with similar cases in your future practice, we will direct this patient to be seated while we give a few moments' time to the consideration of this condition.

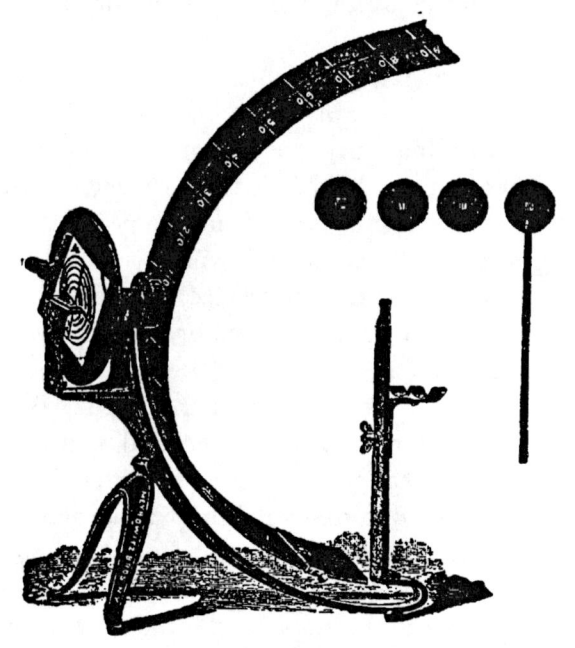

The Perimeter

Tobacco and alcohol cause disturbances of vision with nearly identical symptoms. Either one may produce the disease, but usually both are combined in the same case. It seldom occurs in young persons, the patients usually being over 40 years and with impaired nutrition. We find it more among the poorer classes because poor whisky contains more fusel oil and cheap tobacco more nicotine.

While the ophthalmoscope gives no certain evidence as to the presence of the disease in its early stages, yet as a rule it is not difficult of diagnosis, because of the pronounced subjective symptoms.

CHARACTERISTIC SYMPTOMS

A failure of vision is noticed, most marked in the center of the field. Usually both eyes are affected, but not always to the same extent. The patient sees nothing directly in the line of vision, but objects on either side are seen with more or less distinctness. This interferes with reading and writing, although if the scotoma is not large, while the middle of the sentence is lost the two ends of the line may be seen imperfectly. Patients usually complain more of disturbance of vision when in a bright light. Vision for color also fails in this central scotoma, the perception of red and green being lost.

As we look at our patient he has the appearance of a naturally healthy man, but no one can drink beer and whisky as he does and continue it with impunity. Some men may drink more, and because they have not yet lost their sight or ruined their stomach, they think they are all right. Some men may go through a battle without being shot, and yet no one will argue that war is a safe occupation. So any man that uses alcohol and tobacco to excess is in danger that, sooner or later, some organ will give out.

Is there a scotoma in this case? In order to determine this point we will send for a perimeter and make the test before you. We will use a piece of red card, and we find as it approaches the center of the field, it disappears, and be it remembered this is one of the diagnostic points of a toxic amblyopia, the scotoma for red in the center of the field. We may remark in passing that as the patient begins to recover and the scotoma becomes less noticeable, the patient first sees red as pink, and gradually as improvement continues it becomes redder and darker.

THE PERIMETER

We are sometimes asked by students about to enter on practice, whether a perimeter should be included in their outfit, and we usually answer that we do not consider it necessary, adding a few words of explanation as to the specific uses of the instrument. It consists essentially of an arc, as you see, which can be moved to any position in order to measure any desired meridian, and a small white test object moved in from its outer end until it becomes visible. As this is repeated at a number of different points, say 15° or 20° apart, until the circle is completed, the outlines of the field are indicated. The test object is also moved inward to the center along the different meridians and in this way the presence of any scotomata are detected, and their size and location are mapped out.

The province of the perimeter then is to measure the field of vision and to detect the existence of a scotoma, all of which is included in indirect vision. It has little to do with direct vision, the vision of the yellow spot in which you are particularly interested in fitting glasses, and therefore it is only in exceptional cases that you will even think of it.

If you consider in any case it is desirable to obtain some idea of the extent of the field of vision, it can be determined on the blackboard or any plane surface. A mark is made for the sight to be fixed upon, and a piece of chalk is moved inwards from the extreme limits of the board up to the center. The marks made where it first appears indicate the limits of the field. If it should disappear and reappear, a scotoma would be shown to exist.

Now, it is safe to say the diagnosis has been fairly made out. The patient cannot see well, his vision does not respond to the pinhole nor to any lenses we place before his eyes, proving that it is not an error of refraction ; there is no opacity or any disease of the retina ; there is the scotoma for red in the center of the field ; he is an inordinate user of alcohol and tobacco.

The first step in the treatment is the removal of the cause, and if this man wants to regain his sight, he must abstain from drinking and smoking. In addition we will prescribe strychnia internally, galvanism locally and a good nourishing diet.

INSTRUCTIONS TO PATIENT

-To the patient, "My man, for the sake of your sight and for the benefit of your family that depends upon you, you must cut out your beer, whisky and tobacco. If you will faithfully follow our directions, I think I can safely promise that you will get well. Come back in a week and report."

Note that in this case the prognosis is good, and so it is in any case of toxic amblyopia if patient is seen early and if he will abstain entirely from the poisonous agents. He may even get well if he is willing only to reduce their quantity, but this is not at all certain Persistence in their use is sure to lead to greater impairment of vision or even practical blindness. After complete recovery, the disease may sometimes recur if the appetite for drinking and smoking is uncurbed.

While it scarcely comes within your province to prescribe the administration of a nerve tonic or the application of electricity, yet

I think it is your duty as educated optometrists to be able recognize the existence of this condition, and it will also add greatly to your prestige. Make your examination along the lines I have mentioned, and refer your patient to a medical man for treatment. In this way you establish a reputation for competency with the patient and his physician, and you gain the good will and confidence of both.

Three weeks later we have before us this same patient but a great change has taken place. He has called each week since then, always reporting improvement. He says his eyes have lost that heavy feeling and that they feel good. We test his vision on the card and find that with $+$ 1.50 the acuity equals $\frac{15}{20}$. He says in answer to our question that he has not drank any whisky or beer, but that he has smoked an occasional cigar. This is a very satisfactory improvement and the indications point to complete recovery. We will advise him that his welfare demands continued abstinence, and that he should still keep on with the medicine and the galvanism for some weeks longer.

A Case of Refactive Error Diagnosed as Cataract

[CLINIC No. 7]

The case we have to present for consideration on this occasion is an interesting one, that points a moral not only in the diagnosis between disease and refractive error, but in the technique of the examination of doubtful cases.

Mr. J.W. L., aged thirty-eight years, has been advised to consult us by the optican who fitted him with glasses, on account of cataract in the left eye. This gentleman whom we know to be a competent optometrist, sends this note with patient, in which he says he has fitted right eye with — .50 cyl. axis 180°, and left eye, because of cataract, is fitted indifferently with a weak convex sphere.

We will glance at this patient's eyes, but we look in vain for any noticeable evidence of cataract. A casual examination shows that the pupil of the left eye is just as bright and black as that of the right eye. This at once disproves the diagnosis of cataract, because in this disease, if at all advanced, the pupil becomes white or whitish gray. This grayish appearance is so diagnostic of cataract, that sometimes opticians (and the laity, too) class leucoma (a gray opacity of the cornea) as cataract, a mistake which every intelligent man will want to avoid.

WHAT A CATARACT IS

As cataract is a condition which you will not uncommonly meet in your practice, it is proper that you should have a well-defined idea of just what it is and how you will determine its presence.

Cataract is an opacity of the cystalline lens. It is not a tumor or a growth of any kind, but it is simply a condition in which the crystalline lens loses its transparency and becomes opaque. The lens instead of being transparent as in health, becomes translucent. Remember, this milky appearance is in the lens, not in the cornea; it is behind the iris, not in front of it. When an eye presents this gray color, don't hastily conclude that the patient is suffering from cataract. Examine the eye closely, and if you find this milky appearance located in the corneal tissue, the case is one of leucoma, which is a very different thing from cataract.

If you find the cornea of normal transparancy, which you can determine by simply looking at it, or still better, by oblique illumination, and if the milky appearance occupies the pupil and back of the iris, the trouble is undoubtedly cataract.

In order to demonstrate this, I will ask the patient to be seated on a chair beside the light, and I take a strong convex lens from the trial case and focusing the light on the anterior portion of the eye, we see the cornea and aqueous are transparent and the pupil free from opacity.

If the cataract is complete, it will be readily recognized in the manner I have outlined. In incipient cataract before the whole lens has become affected, the partial opacity is best detected by the ophthalmoscope, using a strong convex lens in the sight hole and holding the instrument about the focal distance of this lens from the eye.

I will rotate a $+ 6$ D. lens into the aperture of my ophthalmoscope, and approaching to about $6\frac{1}{2}$ inches, I find a perfect red reflex, unobstructed by the slightest opacity. This completely disposes of the diagnosis of cataract, but so far we have shed no light on the cause of the impaired vision.

CAUSE OF IMPAIRED VISION

We ask the patient how long his eye has been so bad, and he tells us that it has been so to his certain knowledge for fifteen years, and perhaps longer. It has caused him no great inconvenience, and he assumed that nothing could be done for it.

We place the black rubber disk over right eye and direct patient's attention to the card of test letters. He can see there is something there, but he cannot see that it is a card, much less discern any marks on it.

Will some of you who are following this clinic, tell me what should be the first step in the examination of this eye in order to determine its optical condition? "The use of the pin-hole test," you reply; yes, that is correct. In this way we can quickly determine whether or not any vision is present, and if so, whether it will respond to our test lenses.

The pin-hole disk is placed in trial frame before the eye, and patient begins to name the larger letters: he commences at the top and names the letters on the first four lines. This corresponds to a visual acuity of $\frac{20}{70}$, and we will make a note to this effect in our record book. We now know that there must be a considerable error of refraction present, and that by its measurement we must be able to improve vision at least to an equal amount.

Before commencing our test with trial lenses, we will ask the patient to step over to the ophthalmometer table and see what light this instrument will throw upon the case. First, you must focus the instrument so as to secure the sharpest possible reflection. Then approach the mires until they barely touch each other in the horizontal position. Now, we will slowly rotate the instrument and note the result.

As the instrument is turned the mires at once begin to overlap each other, until as the vertical meridian is reached, there is an overlapping of 1, 2, 3, 4, 5, 6 steps; yes, six full steps. What does this indicate? It shows a corneal astigmatism of 6 D., the curvature of the vertical meridian exceeding the horizontal by that amount; or the horizontal falling below the normal curvature to an equal extent.

Now, how shall we classify this astigmatism? It is either hypermetropic astigmatism in the horizontal meridian, or myopic astigmatism in the vertical meridian, or a combination of both.

THE TRIAL CASE EXAMINATION

We will now proceed with our trial case examination. As the case seems so overwhelmingly one of astigmatism, we will try first a cylinder instead of a sphere, as we usually do; but we will not depart from our rule of always trying convex first. As the astigmatism is evidently of high degree, it seems useless to try a weak lens, and so we will commence with a $+ 3$ D. cylinder, which, according to the indications of the ophthalmometer, we place in trial frame with axis vertical.

The patient says this affords no improvement; in fact, makes vision worse if anything. We feel disappointed because the pinhole showed that vision could be improved, and as the ophthalmometer revealed such a high degree of corneal astigmatism, we assumed the correction of the latter would cause the looked-for improvement in vision.

However, before discarding this lens we will rotate it, and as we do so, we are rather surprised to find that vision begins to improve, reaching its best point when axis is horizontal.

This shows a case of astigmatism against the rule; whereas, the reading of the ophthalmometer showed astigmatism with the rule. But the latter measures only the corneal astigmatism, while the lens corrects the total defect, which is made up of both corneal and lenticular astigmatism. Therefore, we must conclude that the defect in the crystalline lens not only neutralizes that in the cornea, but turns it in the other direction.

LIMITATIONS OF THE OPHTHALMOMETER

Now, I would not for an instant belittle the value of the ophthalmometer; I would not part with mine, and I would advise all of you that can afford it to procure one; but at the same time we must admit that in many cases its readings are not to be depended upon, and this particular case is one of them.

As the patient notices a decided improvement in vision with the convex cylinder, and as he unhesitatingly prefers the axis in the horizontal position, we feel that we have at last struck the right track, which we must follow to its best results.

We therefore gradually increase the strength of the cylinder, producing still greater improvement, until finally we reach $+$ 6 D. cyl. axis 180°, which is the strongest patient will accept, and with which the larger letters on test card are now readable.

As vision is not yet up to the standard set by the pin-hole, we know that we have not reached the proper correction and we must make still further effort. With the thought that perhaps the astigmatism is not fully corrected, we will try a concave cylinder over this convex with its axis at right angles, commencing with a $-$ 1 D. Patient says the addition of this lens affords a marked improvement, and we increase its strength until we reach $-$ 6 D. cyl.

We now have in the trial frame $+$ 6 D. cyl. axis 180° $\bigcirc$ $-$ 6 D. cyl. axis 90°, and we will ask the patient to name letters on lowest line he can see. He says C B L O T G; you will recognize this as the No. 30 line, half of the letters being incorrectly named. We will make note in our record book that with the cross-cylinder correction V. $= \frac{20}{30}$? ? ?, the interrogation points denoting that three of the letters were in doubt. Now this is a pretty good result for an eye that was diagnosed cataractous and given up as practically blind.

But our difficulties are not yet over by any means: it is one thing to correct a high degree of defect, and it is another to get the patient to wear such a lens in a case where the other eye is so nearly normal. And, by the way, we must ascertain its acuteness of vision and its correcting lens, if any. We find that V. $= \frac{20}{20}$ almost, some of the letters being in doubt. After trying convex spheres and cylinders, we find that — .25 D. cyl. axis 20°, affords the greatest acuteness of vision, and makes every letter of the No. 20 line perfectly plain.

We try the two eyes together, each with its correction, but patient says that objects are doubled and the glasses are so confusing that he could not wear them, We must, therefore, reduce the strength of the left lens, and after several trials patient seems to be able to bear

O. D., — .25 cyl. axis 20°
O. S., + 4 cyl, axis 180° $\bigcirc$ — 4 cyl. axis 90°.

and this is the prescription we will give him for constant wear.

For near vision, patient must have full correction, and hence we will give him another prescription, as follows :

O. D., Frosted
O. S., + 6 D. cyl. axis 180° $\frown$ — 6 D. cyl. axis 90°

With this cross cylinder left eye is able to read the finest print, but because of the fact that there has never been a perfect image formed in this eye, the sensibility of the retina is blunted. This we will endeavor to overcome by directing patient to use these glasses "for exercise" about fifteen or twenty minutes each day, selecting print that is not too small.

We will advise this gentleman that he must have patience, that there will be some difficulty in the use of both pairs of glasses, but that the restoration of the sight of this heretofore useless eye is worth all the trouble he may be caused.

Compound Hypermetropic Astigmatism and Presbyopia, Showing when Cylinders May and Should be Omitted From Reading Glasses

[CLINIC NO. 8]

Mrs. Elizabeth B., aged sixty-six years, complains that sight is dim, eyes water, and lids are sometimes agglutinated in the mornings. We ask her if she has ever worn glasses. She replies that she has, and hands us her glasses, which we find by neutralization to be + 2.50 D. for distance, and + 4.50 D. for reading.

We request her to be seated, and on directing her attention to the test card hanging at twenty feet, she tells us she is unable to name any of the letters.

METHOD OF EXAMINATION

In accordance with established custom we commence the examination with convex lenses, starting with + 1 D. This at once improves vision and makes the larger letters readable. We increase + .50 D. at a time until we reach + 2.50 D., with which she can name a few of the letters on the No. 20 line. An additional + .50 D. affords no further improvement, in fact, she thinks makes it worse.

We will now try + .50 D. cyl. over the + 2.50 D. S., which we place in the trial frame in the position customary for convex cylinders, viz., with axis at 90°. This she at once rejects, but before we remove the lens we will try the effect of rotation. As we turn it slowly around she tells us there is some improvement, which reaches its maximum when axis is horizontal. The addition of this cylinder enables her to read all the letters on the No. 20 line, and we will make the following note in our record book:

O. D. with + 2.50 D. S. ◯ + .50 cyl. axis 180°. Vision $\frac{20}{20}$.

We remove the cylinder in order to give her an opportunity to compare vision without it and with it, and she unhestatingly states that she can see much better with it. We will now proceed to examine the other eye in the same way, and we find that it accepts the same combination, which therefore represents the refractive error in each eye. For reading we find, after a few trials, that

+ 5 D. S. is the best with which she gets the clearest vision at fourteen inches. The addition of the cylinders that were accepted for distance affords no improvement, in fact, patient says she can read better without them.

EXCEPTIONAL CASES

Now, why should this cylinder be rejected? If astigmatism exists, is it not proper to correct it in any glasses that may be placed before the eyes, both for reading and distance? This has always been the teaching, and the majority of us in our practice, and perhaps I may say all of us, have followed this custom. I shall not attempt to deny the soundness of this as a general principle, but I will point out to you the class of cases in which we must make an exception, and this lady will serve as our text for that purpose.

In the first place, I want to say that the majority of persons who furnish these exceptions to the general rule are those who have reached or passed middle age; in other words, presbyopes who in addition have both curvature and axial ametropia. Such persons require a sphere and cylinder for general vision, the axis of the latter being horizontal, and in addition a somewhat strong spherical lens for near vision.

There must be a good optical reason for this departure from custom, and we will, therefore, give some consideration to the matter. All of you are aware of the fact that when rays of light pass through a convex sphere they are converged to a point, in such cases it being understood that the light strikes the lens at right angles to its surface. But perhaps you are not so thoroughly familiar with the fact that when the rays of light pass obliquely through the lens, they are not all converged to a point. A tilting of the lens, or an obliquity of the rays, introduces a cylindrical effect, and then the light rays are acted on as if passing through a spherocylindrical lens held with its surface at right angles to the rays.

Let us see if we can prove the truth of this assertion. I will take a — 2 D. lens and place it in the trial frame before the right eye and the opaque disk before the left eye. I place the trial frame on my face and hold a + 2 D. lens before my right eye. The convex lens neutralizes the concave, and my sight is the same as with the naked eye. I look at the card of radiating lines, and they all look equally black and plain.

Now, I will begin to tilt the convex lens vertically, moving the upper part of this lens away from the concave; in other words, tilting the lens on its horizontal axis. As I do so I notice the horizontal lines lose a little in clearness, the vertical remaining as plain as before. The more I tilt the lens the dimmer the horizontal lines become, and not only the horizontal but the oblique lines, until finally they are all blurred and indistinct, except only the vertical, which still retain their original clearness.

PHENOMENA OBSERVED BY TILTING

As the convex lens is tilted around its horizontal axis the converging power of the vertical meridian is increased, while that of the horizontal meridian is but little affected. The refractive power of the lens is now the same as if a convex cylinder had been placed before the sphere with its axis horizontal. Therefore, the rays of light passing through the vertical meridian are brought to a focus sooner than those passing through the horizontal meridian. In other words, the focal distance of the vertical meridian is less than that of the horizontal.

How do these facts apply to the lines as we have just seen them? It will be remembered that the distinctness of the vertical lines depends upon the proper curvature of the horizontal meridian of the eye, while the horizontal lines bear the same relation to the vertical meridian of the eye. Therefore, as the refractive power of the vertical meridian of the lens is increased by tilting, the horizontal lines are brought to a focus before reaching my retina, and are consequently blurred and indistinct, while the vertical lines passing through the horizontal meridian (which is not affected by the tilting) retain their original clearness.

Now, let us see what we have—a convex sphero-cylinder, with the axis of the latter at 180°, or as applied to my eye, an artificial myopic astigmatism with the rule, the excess of curvature being vertically. The next question that occurs, how can it be neutralized or corrected? Obviously, in the sphero-cylindrical combination, the cylindrical effect is neutralized by a concave cylinder with axis in same position, viz., horizontal; while the artificial myopic astigmatism is corrected by the same form of cylinder and in same position.

Now, I will repeat my experiment and make use of this correcting cylinder. As I tilt the convex sphere and notice that the horizontal lines have become indistinct, I hold a — .50 D. cyl. axis 180° in front of the convex lens, and in vertical position—that is, parallel to the concave sphere in the trial frame. This at once corrects the artificial astigmatism, clears the lines and make them all equally black and distinct.

If I tilt the convex lens more I will need stronger and stronger concave cylinders to neutralize the increased convex effect of the vertical meridian and maintain the lines equally clear. As there is nothing like practical demonstration, I will ask you gentlemen to try this experiment for yourselves, so that you can see the added cylindric effect caused by the tilting of the sphere, and the neutralization of the same by a corresponding concave cylinder. Those of you who are not emmetropic will need to put in the trial frame your own correction, and then proceed as I have indicated.

THE LESSON TO BE LEARNED

Now, what is the lesson to be learned from this, and what application can we make of it in our practice?

In the first place every person who wears convex lenses for reading, or writing, or sewing, develops a cylindrical addition to their lenses more or less marked according to their strength and the degree of obliquity of the visual axis. The stronger the lenses and the greater the obliquity the more marked the cylindric effect.

In the second place, in cases of low degree of astigmatism, in which the axis of the correcting cylinder is horizontal, the latter may be omitted from the reading glasses because the same effect is produced by the obliquity of the rays passing through the sphere, and if the cylinder was retained the astigmatism would be over corrected. I will repeat that this does not apply to lenses for constant wear in persons under forty years of age, nor to simple cylinders with axis vertical, but the principle is applicable to persons over forty, who are, therefore, presbyopic, and whose distant vision calls for a convex sphero-cylinder axis horizontal, or even simple cylinders axis horizontal.

This patient is a typical case for the application of this principle. The + 5 D. spherical lenses which she requires for near use will produce a cylindrical effect at least equal to her astigmatism at distance. It has been estimated that a + 1 D. sphere, as usually worn for reading, produces a cylindrical effect of .125 D., and this would afford in our case an added cylinder of .62 D. or an eighth diopter more than we really need. We will, therefore, prescribe for this lady ·

R. and L., + 2.50 D. S. ◯ + .50 D. cyl. axis 180° for distance and general wear, and R. and L., + 5 D. S. for reading, in the knowledge that the latter pair as used for reading, even though simple spheres, will afford just as much cylindric power as those for distance, which are ground both spherical and cylindrical.

I have no hesitation in advising you to follow these suggestions in your practice when you meet cases similar to this one under consideration to-day. If the astigmatism is of low degree the cylinder may be omitted entirely as in this lady's case. If the astigmatism is of higher degree the cylinder may be reduced in accordance with the estimate given you a few moments ago, viz., — .25 D. for each 2 D, of spherical power. You thus under-correct the astigmatism by cylinders, which is then made up by the increased converging power of the vertical meridian of the tilted spherical lens, or rather the lens remaining straight by the oblique direction of the visual axes.

For sake of illustration, if we have an eye that requires + 1 D. S. ◯ + .25 D. cyl. axis 180° for distance, and an additional + 1 D. for reading, we would prescribe + 2 D. S. for reading, omitting the cylinder, but securing the effect of the same by the oblique direction of the visual axes in reading.

If we had a patient whose error of refraction was represented by + 2 D. S. ◯ + 1 D. cyl. axis 180°, with 2 D. of presbyopia in addition, we would prescribe for reading, + 4 D. S. ◯ + .50 axis 180°, correcting half of the astigmatism by the cylinder, and getting the correction of the other half by the increase in power of the vertical meridian of the sphere. You will meet numerous cases in your practice in which you can act on these suggestions with results satisfactory to yourself and your patients.

In my remarks about this case, and in the illustrations I have used, I have referred particularly to convex lenses, but you will see that the same principles apply to concaves, viz., the increase in power of the vertical meridian of a spherical lens when the rays of light pass through it obliquely.

While I have spoken only of horizontal and vertical meridians and axes, I might say that those cases may be included where the meridians are within 10° of these cardinal lines.

POINTS TO BE REMEMBERED

Before closing our clinics for to-day perhaps I ought to say that if the person wears spectacles and the temples are purposely bent angular, so as to give the lenses the proper degree of obliquity, then of course, the conditions are changed and the cylindrical effect of the lenses is not made apparent. Also with the users of eyeglasses, which are so frequently worn with the tops of the lenses tilted forward, these remarks do not apply. In these cases the tilting of the lens places it at right angles to the visual axis and corrects any tendency to cylindrical effect, while in the experiments to which I called your attention a few moments ago, when the visual axis is directed straight forward to distance, the tilting of the lens produces an obliquity of the rays.

I feel that I have presented to-day for your consideration some thoughts that are perhaps new to many of you, and which all of you can think over with profit. These suggestions do not have a wide range of application, as this class of cases forms only a small proportion of the total with which you have to deal, but they will surely be of advantage in properly selected cases, and may enable you to understand your failure to give satisfaction in some cases where the cause of the trouble was obscure, and help you to a more scientific and comfortable correction than would be possible by an optometrist who was ignorant of the principles which I have attempted to make clear.

Presbyopia

[CLINIC No. 9]

You will be called upon in your practice to prescribe glasses for presbyopia more frequently than for any other form of visual defect. Perhaps, however, we should modify this statement or, rather, state it in a different way: Presbyopia undoubtedly heads the list of the forms of eye failure for which glasses are sought, but very many presbyopes purchase their glasses as they do an article of merchandise. They are loath to believe that their eyes are defective. They say that their eyes have always been good until quite recently, and they think perhaps they are strained, or cold has settled in them, and they will be all right in a few days.

But finally the patient begins to realize that the impairment of close vision is permanent, and he drops into a spectacle store and asks for a pair of "rest glasses," or "first glasses." He does not feel the need of a thorough examination of his eyes, and therefore as you are all skilled refractionists, he does not seek your services, and for this reason your record books will perhaps not show as large a proportion of cases of presbyopia as the frequency of this defect would indicate.

Presbyopia is one of the penalties of age. You have learned from your studies that certain changes take place in the eye with the advance of years, all of which tend to make more difficult the act of accommodation.

This presbyopic change is no respecter of persons; it occurs in all classes of society and in the eyes of every person who reaches middle age. The evidence of it is influenced by the refractive condition of the eye, being accentuated in hypermetropia, while in myopia it is less noticeable. This brings out the fact that every person who lives the allotted span of life is called upon to wear glasses at some time or other during his life. The evidence is around us on every hand that many children and young people must wear glasses: but those who perchance escape at this time, find themselves up against it in the "forties," when reading and near work become less and less possible without artificial assistance.

ILLUSTRATIVE CASES OF PRESBYOPIA

Inasmuch as this condition of presbyopia is so common, it may not prove unprofitable for us to devote our clinic hour to-day to its consideration, and illustrate it by several cases that have presented themselves for examination.

Mr. G. P. L., aged forty-four years, complains of headache after reading, or any close use of eyes.

When a person of this age complains of difficulty in close use of eyes, we at once suspect presbyopia. But as educated optometrists, we have no right to proceed to prescribe glasses on this assumption, but we must prove or disprove it.

In the first place we determine the acuteness of vision. We ask the patient to read the lowest line of letters on the test card, and he names every one on the No. 20 line. This being normal vision excludes myopia, and in order to determine the possibility of hypermetropia, or slight hypermetropic astigmatism, we try a + .50 sphere and a + .50 cylinder, rotating the latter through the different meridians. Both of these being rejected, we are justified in concluding that the refraction is emmetropic.

We now hand the patient the reading card, and ask him if he can read the smallest type at the top of the card. He replies that he can, at the same time moving the card away from his eyes. We request him to bring the card nearer, as close to his eyes as it is possible for him to make out the words even though he feels it a great strain, and on measuring this distance we find it to be nine inches, which represents the near point in this case, thus indicating a deficiency of accommodation.

The question occurs as to what is the standard by which the accommodation must be gaged? The least amount of accommodation which an eye should possess in order to use the eyes for close vision without strain, is 5 D., and therefore in this patient's case there is a deficiency of accommodation.

Now we have the essential features of this case as follows·

Refraction, normal.
Accommodation, deficient.
Age, forty-four years.

The existence of these three factors in any one case proves presbyopia. If any one is missing, it cannot be simple presbyopia; and when all are present, it cannot be anything else.

You will doubtless meet with cases in which the refraction is apparently normal, but with deficiency of accommodation, in persons who are less than forty years of age. Such patients cannot be presbyopic, although they are often wrongly classed as such. If the refraction is carefully examined (as, for instance, by the fogging method and by skiascopy) it will be found that the refraction is hypermetropic, showing itself as a so-called early presbyopia.

Many optometrists in their examination of cases, assume that the refraction is normal when they find the acuteness of vision to equal $\frac{20}{20}$. But you must not too hastily jump to such a conclusion, else you will oftentimes overlook hypermetropia and hypermetropic astigmatism.

WHAT GLASSES TO PRESCRIBE

We are safe then in diagnosing this case as one of presbyopia, and now the question occurs as to what glass should be prescribed? Theoretically, this is a simple matter: the eye should possess 5 D. of accommodation, this man has only 4.50 D. (as shown by the near point of nine inches), and the deficiency is therefore .50 D., which is made up by a convex lens of like amount.

We will ask this man if he has ever worn glasses. He replies that he got a pair about six months ago, but they "pulled his eyes and he hasn't worn them much." We ask to see the glasses, and on neutralizing them find them to be $+$ 1 D.

According to the slight impairment of the accommodation shown by our examination, you will say that these glasses are too strong, and this has been proved by our patient's experience with them.

In the commencement of presbyopia it is not well to give glasses too strong at first. It is a fact that should guide you in your selection of glasses, that there is much more complaint from glasses too strong than from those too weak.

We will therefore order $+$ .50 D. lenses for this man for close work only, with every confidence that they will prove satisfactory.

The next patient is S. W. B., aged forty-five years, who complains that for the past year he has had difficulty in reading, especially at nights.

We find the acuteness of vision in each eye is $\frac{20}{20}$, and that convex lenses are positively rejected. This proves the refraction to be emmetropic.

We give him the card of small type, and he cannot read the first or the second size. He can read only the third, and that no closer than fifteen inches.

We have now all the essentials for a diagnosis of presbyopia: emmetropic eyes, deficient accommodation and age over forty.

COMMON SENSE VERSUS THEORY

Now what glasses shall we order in this case? The near point of fifteen inches, and that with no smaller type than No. 3, indicates an amplitude of accommodation of not more than 2.50 D., which is just one-half of what a person should possess for comfortable reading. In this case, therefore, a glass of + 2.50 D. is theoretically required to bring the accommodation up to the standard of 5 D. Shall we prescribe so strong a glass for a person who has heretofore never worn glasses? No, it would scarcely be proper. We cannot tie ourselves down to any hard and fast rules. They must be modified to suit individual cases, and we must make use of common sense and good judgment all the time.

With the knowledge that strong glasses cause more discomfort than weak ones, we would not dare in this case to give first glasses of + 2.50 D.

Without any special rule to guide us in the selection of glasses in such conditions, but relying on our experience in similar cases, we will order + 1 D. lenses for close use to begin with. This man has been reading up to the present time without any help at all; if we assist him to the extent of + 1 D. it will be pleasantly received. But if we offer him too much help, if we over-assist him, discomfort will result and the glasses will be rejected.

Our third and last case in to-day's clinic is I. B. J., aged forty-nine years, who acknowledges to some difficulty in reading.

The acuteness of vision is $\frac{20}{20}$, and all convex lenses are rejected.

Of course, you will notice we are testing his distant vision with the lenses, and when we say all convex lenses are rejected it is understood at twenty feet, as the examination of the condition of the refraction must be made with rays of light that are as nearly as possible parallel.

On testing the near vision we find that patient cannot see the No. 1 type, but he reads No. 2 as close as twelve inches.

We now have the three factors on which the diagnosis of presbyopia rests.

What glasses shall we give this man? His amplitude of accommodation is 3.25 D., as shown by the near point of twelve inches, therefore he needs + 1.75 D. glasses to bring his accommodative power up to the standard of 5 D. But is it wise to give him glasses as strong as this? Remember he has never worn glasses, and if we give him + 1.75 D. the crutch will probably be too high for comfort. He has been so long accustomed to stooping over that he cannot bear being raised to the upright standard all at once. We will therefore prescribe + 1 D. for close use in the confidence that they will afford more satisfaction than the stronger lenses which are seemingly indicated.

SUMMARY OF RESULTS

Perhaps you will have noticed that all three of the cases have been handled differently.

In the first case (G. P. L.) we gave the full correction as indicated by the rule, viz., to supply such glasses that, added to the amplitude of accommodation, will equal 5 D. We were justified in this by the age of the patient, and that even the full correction was a comparatively weak glass, as weak as would be of any special value. This man's accommodation is but little impaired, but he shows wisdom in seeking optical help as soon as he feels the need of it, instead of struggling for awhile against the inevitable as the other two have done.

In the second case (S. W. B.) we gave less than half the indicated correction. We were guided in this selection not only by the fact of these being the first glasses, but because of the age of the patient, being only five and forty. It is possible his glasses may need changing in a year, but in spite of this it is not well to insist on a stronger glass at first.

In the third csse (I. B. J.) we gave more than half the indicated correction because the patient was so near to the half century mark, thus justifying a greater proportional increase in the strength of the glasses.

These cases were all simple presbyopia, all coming for their first glasses, and yet there were points of difference in each, thus emphasizing the fact that in optometry there are no hard and fast rules to be blindly followed even in cases suffering from the same form of optical defect.

We have made a somewhat hasty examination of these cases, but the history of them led us to believe that they were simply presbyopic, and as they were all in the "forties," we relied on a subjective examination alone, investigating only the condition of the accommodation and the refraction. The first being deficient, and the second normal, the diagnosis could be none other than presbyopia.

But we would earnestly advise you to go more thoroughly into the examination of your cases. You claim to be skilled optometrists, and you want to establish a reputation as such, and therefore in the examination of all cases that seek your services, you should investigate every function of the eyes and the condition of all their parts that are accessible to your trained skill.

After determining the state of the refraction, you should look into the muscular equilibrium and thus detect any heterophoria that may be present, either at twenty feet or at reading distance. You should use your ophthalmoscope to determine the integrity of the refracting media and of the nerve head and retina. Also your retinoscope to verify the condition of the refraction.

In closing I want to say that presbyopia is not a well-defined departure form normal shape or structure (as are the several refractive errors), but rather an impairment of the function of accommodation that comes on so gradually that to a certain extent the eyes adapt themselves to it. Even though the refraction is the same in two persons of the same age, the accommodation is apt to vary, and hence also the measurable amount of presbyopia.

One objection to stronger glasses which we have failed to mention is that they diminish the range of accommodation and tie the patient down too close to his book or work. This imposes a strain on both the accommodation and the convergence. Diminish the strength of the glasses by .25 D. or .50 D., and you will greatly extend the range of accommodation, and at the same time make patient more comfortable in the use of his eyes.

Myopia

[Clinic No. 10]

As a cursory glance through the patients before us shows several cases of myopia, we will devote our attention on this occasion to the consideration of this important refractive error.

This defect is one that you do not meet every day; in fact, the statement is made that scarcely more than one per cent. of all eyes are affected with simple myopia. And yet on account of its tendency to progress and the damage to ocular tissues in high myopia, it calls for the greatest skill and judgment in its management.

J. M., twenty-one years of age. Cigarmaker by occupation. The history he gives us is as follows: for the past three years he has noticed that he was short-sighted. Complains of inability to see across the street and even to recognize persons when passing them on the same side of the street, except by their voice. Usually no trouble in ordinary reading, but pain and lachrymation when reading music, more noticeable at night.

This history indicates a condition of greatly impaired vision. The patient himself says he is short-sighted, but we must not accept the uncorroborated statement of any patient, nor must we even allow it to influence our judgment, else we will often be led astray. Many a patient has come to us and told us he was near-sighted, when an examination showed him to be hypermetropic, of such high degree that distant vision was impaired and the book was held close to eyes in order to take advantage of the larger visual angle. Therefore we must approach the examination of a case with unbiased judgment and proceed in accordance with our usual methods.

IMPAIRED VISUAL ACUITY

In trying to determine the visual acuteness in this case, we find that it is very greatly impaired. The patient is scarcely able to see the test card, much less to distinguish any letters on it. We ask him to approach the card, and it is not until he gets within four or five feet that he is able to name the largest letter at the top of the card. The visual acuity is, therefore, expressed as $\frac{5}{200}$.

When we hand him the reading test card you see he at once brings it very close to his eyes, and on measuring the distances we find that his near-point is three inches, and his far-point six inches, which shows a very limited range of accommodation.

The examination thus far has shown a lessened visual acuity and a restricted far-point, both of which conditions are symptomatic of myopia, but they are also found in high hypermetropia. Now, how are we going to decide which defect is present?

The retinoscope will determine this question for us. The reflex will be dull and the movements slow in either case. If hypermetropia is present, the movement is seen to be "with," and convex lens will brighten the reflex and make movement faster and more noticeable. In myopia the movement is "against," and a concave lens is needed to clear the reflex and the movement.

The ophthalmoscope might also be used; the direct method is preferable, and if hypermetropia is present, the disk is small and the picture of the fundus is improved by convex lenses.

In myopia the optic disk is large and indistinct, and a concave lens is needed to reduce it more to its normal size and make it clearer.

But doubtless some of the members of the class are not familiar with these objective methods of examination, and for their benefit we will adhere to the subjective form of examination by means of the trial case.

In accordance with our rule we will try convex lenses first, and as the symptoms indicate a high degree of defect, it is useless to commence with such weak lenses as .25 D. or .50 D. We will try a + 2 D. Patient says it makes vision decidedly worse. You will notice that this young man is not seated at our usual distance of twenty feet, but that I have brought him closer and placed his chair at about six feet. In cases like this, where the vision is so greatly impaired, the retinal images produced by objects at a distance of twenty feet are so much blurred that patient is scarcely able to determine whether a certain lens makes vision better or worse, and hence the tests at the usual distance are unsatisfactory.

When, however, the patient is seated where we have placed him, the vision is sufficiently good to enable the top letter to be deciphered, and hence the effect of certain lenses can be much better determined. If the patient was seated at the usual distance, the vision is so very much blurred that he would scarcely be able to tell me whether the addition of the convex lens made it any worse or not.

We have allowed this $+\,2$ D. lens to remain before the eye for a moment, because if hypermetropia is present it will encourage a relaxation of the accommodation and vision gradually improves as the eye adapts itself to the glass. But in this case there is no such improvement, and hence we remove the lens and replace it with a $-\,2$ D. The change in the expression of the patient's face tells us at once that this lens is more acceptable, and in answer to our question he says the letter E is much sharper and blacker and he can almost determine the two letters on the second line.

MYOPIA INDICATED

We have now proven that the refraction of this young man's eyes is myopic, and this indicates the direction along which our examination must be made. We now ask our patient to return to the chair at the customary twenty feet distance, and we will see what information can be gained from the pin-hole disk. We ask the patient to hold it close in front of each eye while the other is covered, and after a little hesitation he names these letters in answer to our inquiry as to the lowest line he can read, C B L C T O, which you will recognize as the No. 30 line, which properly reads O E L C T G. This test proves that the impaired vision is due to a refractive error and indicates the extent to which it can be improved, and as our previous test showed the refraction to be myopic, we now have a pretty fair understanding of the case and all that remains is to measure the amount of the myopia.

We test each eye separately, gradually increasing the strength of the concave lenses until we reach $-\,6$ D., with which V. $=\frac{20}{30}$. A $-\,6.50$ D. lens affords the same acuteness of vision, but makes the letters smaller. A $-\,.50$ D. cyl. axis 180°, placed in front of the $-\,6$ D. is accepted as an improvement and enables patient to guess that the first letter on the No. 20 line is an A, all other letters being blurred and illegible.

Now that we have measured the error of refraction, what glasses shall we order? It would seem like a simple matter to give this young man a pair of $-\,6$ D. S. ⌒ $-\,.50$ D. cyl. axis 180° and send him on his way rejoicing, but I want to say to you that the prescribing of glasses in myopia is a serious matter and should receive the most careful consideration from every point of view.

This case may properly be classed as one of high myopia, in which two pairs of glasses should be ordered, one pair for distant vision and the other pair for close use.

PRESCRIBING GLASSES IN MYOPIA

This young man has never worn glasses and his myopia has undoubtedly been increasing from year to year as a result of the neglect. On account of the myopic condition of his eyes, by means of which they are adapted for near vision, but little effort of accommodation has been required. As a consequence, his ciliary muscle is weak and deficient in sphincter fibers, and, in fact, may be considered as somewhat atrophied. If we should give him a full correction with instructions to wear the glasses constantly, both far and near, a heavy tax would be placed upon the ciliary muscle, with one of two results as follows: the muscle would be forced into unaccustomed activity, making the patient very uncomfortable and with all the symptoms of marked asthenopia. Or the muscle would be unable to respond to the call upon it, and as a consequence the rays of light could not be focused upon the retina and vision would be indistinct, especially for near objects, in fact reading would be an impossibility.

These facts lead to the rule that was long ago adopted to govern the prescribing of glasses in myopia, viz., that they should be the weakest possible consistent with a fair visual acuity, and that in the higher degrees even these must be modified for close vision.

For the reasons just mentioned we do not think it advisable to order glasses stronger than -5 D. for distant vision, and -3 D. for reading and work, omitting the cylinder as of no very special advantage.

A myopic eye is really a "sick" eye, and the prescribing of glasses is only a part of the general treatment which such eyes should receive. Without proper precautions glasses alone may do more harm than good. When myopia is once established, its constant tendency is to increase not only to the great impairment of vision, but to actual damage to the tissues of the eye. If we had seen this young man ten years ago, when perhaps his myopia was less than half it is to-day, we feel sure we could have checked the progress of the defect and sent him through life with a much better pair of eyes than he has to-day. But "it's better late than never," and hence we will give him some advice in addition to his glasses.

BE CAUTIOUS IN USING THE GLASSES

He should be careful in reading and working to have a good steady clear light: coming, if possible, from the left side but never from in front.

If it is possible for this young man to do so, we would advise him to change his occupation for one in which he would not be so closely confined. If not, his hours of work should be restricted and he should arrange for a great deal of out-door exercise. After working all day, he should do no reading at night or very little.

While at work at his bench he should sit erect and incline his head as little as possible, as otherwise a congestion of the ocular tissues is favored, which is a cause of the origin and continuance of a myopia. This keeps his work at a proper distance from his eyes, and the same precaution should be observed in reading and writing.

We would advise him to abandon his music to a great extent, and in what little reading he does to select clear print of good size.

Persons engaged in sedentary occupations as he is (cigar-making, you will remember), are apt to be of a constipated habit. He should endeavor to correct this tendency and pay every attention to the condition of his health.

As I have laid so much stress on the seriousness of myopia and the necessity for care in the use of the eyes and skill in the adjustment of glasses, it may be well to say a word as to the causes that produce this condition.

DEFINITION OF MYOPIA

Myopia, as you all know from your previous studies, is that condition of refraction in which the rays of light come to a focus in front of the retina. This may be caused by an increase of refracting power of some of the media, as a shortening of the radius of curvature of the cornea or an increase in the convexity of the crystalline lens, the latter being produced by spasm of the accommodation, at first only temporarily, but which later develops into a permanent condition.

However, our usual understanding of a myopic eye is one in which there is an elongation of the ball itself in the antero-posterior direction.

While heredity is looked upon as a predisposing cause, you must not think that the babe is born necessarily with myopic or elongated eyes. Only the predisposition to myopia is inherited, which means that the coats of the eye are weak and elongate from an amount of overuse or strain that would have no effect upon the normal eye.

In this connection we wish to impress upon you the importance of being on your guard not to mistake a case of false myopia (which is really hypermetropia over-corrected by spasm of the accommodation) for real myopia, and we hope none of you will ever be guilty of putting concave lenses on hypermetropic eyes. This is an error that can be easily made, and in our experience we have seen many examples of it.

HYPERMETROPIA SIMULATING MYOPIA

The hypermetropic child simulates myopia, because he holds his book very close. He does this in order to obtain a larger retinal image; the image is not clear because it is composed mostly of diffusion circles, but what it lacks in clearness it makes up in size.

When such a child is brought to the optometrist, the statement is usually made by the parents or by the child himself that he is near-sighted. Such statement is made in all innocence, because of the near position of the book. But if you accept such statement as indicating myopia, which has often been done, you are making the first step in the path of grievous error. And if you are led to regard the case as one of myopia, what is more natural than that you should commence your test with concave lenses; this means a great many steps in the same path of error, because such concave lenses will be accepted by the patient and then prescribed by you. They are accepted because the diminishing effect of the concave lenses serves to neutralize the excessive refraction caused by the spasm of accommodation. But the wearing of such concave lenses spurs the ciliary muscle to still greater contraction, with accompanying symptoms of asthenopia, which will soon transform this case of simulated into real myopia, and all because you were too lazy to make the proper examination yourself or too ignorant to know how to do it, but were satisfied to accept the patient's diagnosis without a question.

If now, on the other hand, you ascertain for yourself by the various tests with which you are all familiar, that the refraction is really hypermetropic and you prescribe the proper convex lenses, the after history of the case will be entirely different. The convex lenses will relieve the strain on the accommodation, will form clear and well-defined images on the retina in place of diffusion circles, the necessity for holding the book close will be removed, and the eyes will be no longer directed myopic-ward, but will rather be turned towards emmetropia.

A WORD OF CAUTION

Therefore, when a child is brought to you with a history of near-sightedness, based on holding the book close, do not fall into any such trap as I have mentioned. Remember that very few children have myopia, that the prevailing error of refraction in childhood is hypermetropia. Therefore, in any child, no matter what the symptoms, always suspect hypermetropia, and make your examinations with this thought constantly in your mind.

I feel that I cannot too strongly emphasize the importance of this matter. The examination of a child's eyes, when the tissues are soft and yielding and liable to give way under strain, imposes upon us a much greater responsibility than in the case of an adult where the coats of an eye are firm and unyielding and able to resist the tendency to elongation.

It is unnecessary to add that at all times the cases of children should be given the most scrupulous care by the refractionist. The child's life and preparation therefor are still before it, and its future may, to some extent, be influenced for good or the reverse by proper or improper treatment of its eye defect at the hands of the optician.

Monocular Vision

[CLINIC NO. 11]

Among the patients who have applied to us for treatment recently were three whose visual defect was monocular vision, and the thought occurred to us that it would not be unprofitable to bring them together and present them as the subject of this clinic.

Miss E. V. H. is thirty-five years of age. She tells us she had convergent strabismus in youth, at which time a tenotomy was performed. She claims now that her eyes ache continuously and have been doing so for the past six months.

As we look at this lady's eyes we notice at once that there is still a condition of convergent strabismus, the right eye being the deviating eye.

An examination of the acuteness of vision results as follows ·

R. E., unable to see any letters on the test card.

L. E., $\frac{20}{20}$? ? ?. You will notice that she reads the No. 20 line, but that she mistakes three of the letters on the line.

We now proceed to make a more careful examination. We cover the left eye and lead the patient toward the card asking her every step or two she makes whether she is able to see any letters on the card. It is not until she gets within five feet of the card that she is able to distinguish the largest letter, and we have, therefore, made a record of the acuteness of vision as $\frac{5}{200}$.

THE PIN-HOLE TEST

You will, perhaps, recall our method of precedure in cases of extremely defective vision, which we have so often described, and that is to determine the possibility of improvement in vision by means of the pin-hole disk. If the pin-hole improves vision, we can expect an equal (and sometimes a greater) improvement by means of the lenses. While, on the other hand, if the pin-hole fails to afford an improvement, you will know the case is hopeless and no benefit is to be expected from glasses.

We will, therefore, take the pin-hole disk from the trial case, and still keeping the left eye covered, direct the patient to hold the disk close in front of the sight of the right eye. The sight is so greatly impaired that patient finds considerable difficulty in getting the disk in proper position. At first she says everything is black because she is trying to look through the opaque part of the disk. But finally, after several trials, she succeeds in placing it in the line of vision, but the result is nil. We have allowed the patient to remain at the distance at which she could first see the large letter, but there is so much loss of illumination and so few rays of light enter the eye, that even the large letter at the top of the card now is scarcely legible. This proves that the impaired vision is not due to an error of refraction and that no improvement is to be expected from lenses. We will now make an ophthalmoscopic examination. We find that all the media are clear and that there are no marked changes at fundus of the eye. The retina and optic disk are paler than normal and the details are but dimly seen. We diagnose the condition as amblyopia.

Now a word in passing as to the cause of this amblyopic condition of the retina. It is undoubtedly due to non-use and is present to a greater or less extent in nearly every case of strabismus. I can assure you, however, that it is a preventable condition, but is allowed to become amblyopic from neglect, from failure to consult a specialist who could have pointed out the dangers and indicated the proper methods to be pursued to preserve vision.

PROBABLE HISTORY OF THE CASE

It is more than likely that this lady started life with an equal amount of vision in each eve. That her eyes were hypermetropic. That at the age of three or four (or perhaps a year or two later) a convergent strabismus began to develop. That this excessive convergence was caused by the extra accommodation required to overcome hypermetropia when she first commenced to use her eyes for near vision. That the tenotomy was probably not performed early enough and that it failed to restore binocular vision.

We said that this amblyopia was preventable. When this operation was performed in youth, the vision of each eye should have been carefully examined. If that of the right eye was found deficient, a special effort should have been made to improve it. This can be accomplished by covering the good eye and compelling the defective eye to perform the act of vision, and continuing this procedure until the vision of the two eyes is nearly or quite equalized. Then binocular vision can be cultivated, perhaps by the assistance of prisms.

In the consideration of cases of this kind, there are two things of which you may be assured

1. That when an eye fails to participate in the act of vision, as in the deviating eye in strabismus, it loses its sharpness of sight and becomes amblyopic.

2. That if such an eye is compelled to perform its function, under persistent and intelligent direction, the sight can be very greatly improved and in the majority of cases restored to normal. We have been able to accomplish this result in our work here in a great many cases, the most favorable for improvement, of course, being those who are young and where the amblyopia is not of long standing. But the older cases are not, by any means, hopeless, as has been proven several times to my knowledge, where an accident in adult life has destroyed the sight of the good eye, the amblyopic eye which was formerly useless now begins to assume the work of vision and soon develops into a useful organ. But prevention is better than cure always, and therefore I would impress upon you the importance of recognizing these cases in early childhood in order that you may be in position to give the proper advise to prevent a failure of vision which is otherwise inevitable.

In this case the right eye is practically useless as an organ of vision. We call this condition monocular vision, which literally means one-eye vision. The time for making an effort to improve vision has long since passed, because it would require more time, patience and perseverance than a patient is able or willing to give to it. Therefore, in this case, our efforts must be directed to preserving and conserving the sight of the other eye.

We find the acuteness of vision of this eye $\frac{20}{20}$ partly. Before commencing the test with trial lenses, we will ask to see the glasses patient has been wearing, and on neutralizing them find that the left lens is $+ .75$ D., but which patient says is unsatisfactory.

We place a $+ .50$ D. sphere in the trial frame. Patient says this makes no difference, vision is just the same with it as without it. We substitute a $+ .50$ cylinder, which we place with axis at $90°$. This we are told is no better, but before removing it, as is our usual custom, we will rotate it. As the axis turns toward patient's left, she says it makes vision notably worse. We now rotate in the opposite direction and patient tells us this begins to improve vision. After a few trials we find that the proper position for the axis is at 135°, when vision is at its best and the No. 20 line is quite legible.

WHAT CONVERGENT STRABISMUS INDICATES

The presence of the convergent strabismus leads us to believe that there is more of a hypermetropic element in the case than is indicated by the convex cylinder that has just been accepted. In order to prove or disprove this supposition, we place convex spheres in the trial frame in connection with the cylinder. We commence with + .50, with which patient says vision is the same; + .75 and + 1 are accepted in turn, but when + 1.25 is placed before the eye, she says she cannot see the letters so well. We therefore replace it with the + 1, which in combination with the cylinder affords a vision $\frac{20}{20}$, clearly and comfortably. This compound lens then represents the manifest error of refraction, which is a compound hypermetropic astigmatism.

We have worked out the correction in this case pretty carefully, and there is scarcely any doubt that the formula mentioned would be the proper one to prescribe. Doubtless many of you think we have already spent too much time in the examination of this case, and would feel satisfied to dismiss it without any further investigation. But we hope there are some among you who feel that an educated optometrist should not rest content with a simple test case examination and correction. And this is the sentiment I want to encourage, in the hope that it will spread until all of you are of the same mind. There is no question that the man who is confined to one method of examination and knows no other, is not entitled to a front seat in the profession of optometry, nor will he be able to hold his own when he comes in competition with more versatile refractionists.

FOGGING AND RETINOSCOPE TESTS

Therefore, we will employ two other methods of examination, in order to determine if the result obtained from the first method is correct.

First, the fogging method. We place a + 5 D. lens in the trial frame, which of course, greatly fogs the vision. We reduce with concaves until we have + 2.25 value in front of the eye with which the No. 20 line is partly legible. We then direct patient's attention to the card of radiating lines, and in answer to our question as to the difference in their distinctness, she says the lines from 30° to 60° are dull as compared with the others. With this as a clue, we place a — .50 cylinder with its axis oblique in front of the convex sphere. This at once clears up the lines, and on directing her attention to the letters, we find that the No. 20 line has been made much clearer by the addition of the cylinder. By transposing this formula, the result of the fogging test is + 1.75 S. ◯ + .50 cyl. axis 135°.

We will now measure the refraction by the retinoscope. As we cause the shadow to move across the naked pupil, we see at once the movement is "with." We place a + 1 D. in the trial frame and the movement is still with ; we replace this lens with a + 2 D., with the same result. We increase to + 2.50 D. with which the movement is neutralized in some of the meridians. On careful observation we find that the meridian in which there is the most decided "with" movement remaining, is the 45th. This calls for a further increase in this meridian, and we find that a + 3 D. "kills" the movement. Now, what has the retinoscope given us? Perhaps we can make the retinoscope findings clearer to you by a diagram on the blackboard

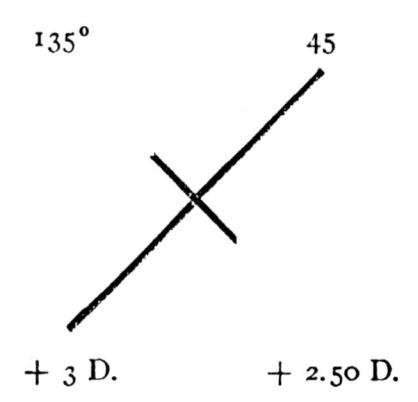

```
      135°              45
            \          /
             \        /
              \      /
               \    /
                \  /
                 \/
                 /\
                /  \
               /    \
              /      \
             /        \
            /          \
      + 3 D.         + 2.50 D.
```

After making allowance for the one meter at which test is made, the result is as follows : The refraction of the 45th meridian is + 2 D., and of the 135th, + 1.50 D., which would call for the following correction : + 1.50 D. ◯ + .50 D. cyl. axis 135°.

A comparison of the results obtained by these three methods, although at first sight apparently disagreeing, really shows no discrepancy.

The test case measures the manifest error, while the retinoscope and the fogging method bring out the latent error, the latter, in this case, developing more of the latent defect than does the retinoscope.

ANOTHER ILLUSTRATIVE CASE

Mrs. S. W. K., aged sixty-seven years. We have been unable to get much of a history from this old lady. All she has to say is, that she thinks her glasses need changing.

R. E., vision = $\frac{20}{80}$. With + 2 S. ○ + .50 cyl. axis 90° = $\frac{20}{30}$.

L. E. No letters can be seen on the test card. The use of the pin-hole disk gives a negative result, and, therefore, it is useless to waste our time in trying lenses. For reading we find that + 5.50 affords the best vision.

An examination of her old glasses shows them to be + 2 D., for distance, and + 5 for reading, and we are, therefore, justified in changing them to correspond to our examination. Our special interest in this case, at this time, lies in the fact that it is one of this series of monocular vision.

H. R., aged twenty years. Telegraph operator. Left eye diverges, twitches and waters.

R. E., vision = $\frac{20}{40}$??. With + 2 cyl. axis 110° ○ + .50 S. = $\frac{20}{20}$.

L. E., vision reduced to counting fingers. Pin-hole disk gives a negative result. The ophthalmoscope shows an atrophy of the retina, which is very white and with but few blood vessels and these like small threads.

This little series of cases all present the one feature of monocular vision. In two of the cases the contributing cause is a strabismus; in one case convergent, in the other case divergent. In the other case there is no deviation of the visual axis, and hence the cause of the loss of vision is in doubt.

In all these cases the one eye is beyond our help, and our duty is confined to a careful correction of the other eye, in order to relieve it from undue strain and preserve it in as good condition as possible. While the loss of the sight of an eye may be considered quite a serious matter, yet none of these patients seem to be inconvenienced by the sight being confined to one eye, because they have been so long accustomed to monocular vision that it has become a second nature to them.

Mixed Astigmatism

[CLINIC NO. 12]

This young man (H. L. P.) is twenty-two years of age, and has just graduated as a civil engineer. He complains of drowsiness after reading, and feels that he can continue only by forcing himself to do so. You see he is already wearing glasses, which he tells us he has had for six years. I will ask one of you to neutralize these lenses and tell us what they are.

THE NEUTRALIZATION OF LENSES

In the meantime, it will not be out of place to give you a few points on the neutralization of lenses. In the first place, you should ascertain whether the lens is convex or concave. This is determined by the apparent motion of objects seen through the lens. If the motion is opposite, the lens is convex; if in the same direction, the lens is concave.

The next point to be determined is whether the lens is a sphere, a cylinder or a combination of both. For this purpose the lens is rotated while we look through it at a straight line, as for instance the edge of the window sash or of a picture frame. If this edge breaks as the lens is turned, there is a cylindrical element in it. If there is no break, it is a simple sphere. In the first case, we must next determine whether the cylinder is plano or compound; if we can move the lens in any one direction without causing movement of the object looked at, it is a plano-cylinder; but if there is movement in every meridian, in some meridians more marked than others, it is a sphero-cylinder.

Our young friend is now ready to report on this lens, and he tells us it is a — 2 D. cylinder with the axis at 180°. Now let us look at the lens and see how he arrived at this conclusion. As we hold the lens about ten inches in front of our eye and look through it at the window sash while we rotate it, we see a movement of that part of the sash seen through the lens, showing that it is a cylinder. As we move the lens from side to side there is no motion; as we move it up and down, there is a decided "with" movement, thus showing that it is a plano-cylinder and that it is concave.

In order to neutralize it we take a convex cylinder from the trial case, selecting + 1.50 D., and placing over the lens with its axis horizontal, move it sideways, no motion; move it up and down, still a with movement. This convex cylinder is therefore not strong enough to neutralize the concave. We try a + 2.50 D. cylinder in the same position, and now the up-and-down movement is opposite, showing this lens to be too strong. We now select a + 2. D. cylinder, and placing over the lens with the axis horizontal, we find there is perfect neutralization, no movement in any direction, and as this is accomplished by a + 2 D. cylinder axis 180°, we know that the value of the lens is − 2 D. cylinder axis 180°

EXAMINATION OF THE PATIENT

We will now proceed to the examination of our patient, and we find the acuteness of vision in each eye is $\frac{20}{60}$, while the range of accommodation with .50 D. type is from 4″ to 28″.

Bearing in mind the fact which I have so often tried to impress upon your minds that hypermetropic astigmatism is likely to be mistaken for myopic, we will not allow ourselves to be misled by the fact that patient has been wearing concave cylinders, but we will follow our invariable rule of trying convexes first.

A + .50 D. sphere is positively rejected. We now try a + .50 D. cylinder, which in accordance with the usual custom, is placed axis at 90°. This is accepted as affording an improvement in vision, + .75 D. cylinder and + 1 D. cylinder are each accepted as better than the previous ones, but + 1.25 D. cylinder is rejected. We rotate the cylinder first to the right, then to the left, in each case blurring the letters, so that we conclude that the proper position for the axis of this 1 D. cylinder is at 90°, which indicates a hypermetropia of 1 D. in the horizontal meridian.

Now while this convex cylinder has afforded a marked improvement in vision, it has failed to raise it to normal, and, therefore, we will try the effect of a concave cylinder, commencing with − .50, which as a matter of course we place with axis at 180°. This at once sharpens and blackens the letters, and we keep on increasing the concave cylinders until we reach − 2 D., when we find the acuteness of vision is full $\frac{20}{20}$. This lens represents a myopia of 2 D. in the vertical meridian.

By this simple method of testing, we have arrived at the conclusion that this case is one of mixed astigmatism.

SOURCE OF ERROR TO BEWARE OF

The optometrist who fitted his present glasses entirely overlooked his hypermetropic meridian, and corrected only his myopic meridian. If we had commenced our test with concave cylinders, they would have been accepted and would probably have raised vision to normal, and we would have fallen into the same error as the previous fitter. This only serves to emphasize the importance in our test case examinations of commencing with convexes, and endeavoring to have them accepted, even though the patient is at first inclined to say they are "no better."

Mixed astigmatism is not so uncommon as it might seem to be. The statement is made that six persons out of every hundred are affected with this form of error of refraction. It is a combination of simple hypermetropic astigmatism with simple myopic astigmatism, with the axis of the correcting cylinders exactly at right angles to each other.

The correction as we found it in this case is $+ 1$ D. cylinder axis $90°$ $\bigcirc$ $- 2$ D. cylinder axis $180°$. This you will recognize as a case of astigmatism "with the rule" the vertical or myopic meridian focuses parallel rays in front of the retina, while the focus for parallel rays of the horizontal or hypermetropic meridian is behind the retina.

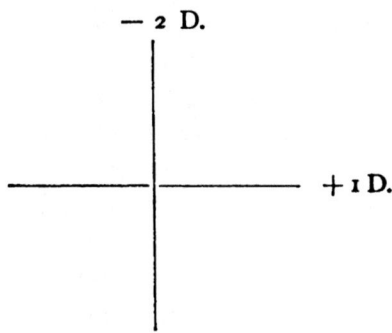

$$-2\ D.$$
$$+1\ D.$$

This little diagram which I have placed on the blackboard will serve to fix in your minds the refraction of the two principal meridians, which corresponds to the correcting cross cylinder, which we also place on the blackboard.

This cross cylinder may be transposed to a sphero-cylinder in accordance with the following rule:

Take either one of the cylinders for the sphere, and the sum of the two for the cylinder, retaining the sign and axis of the latter.

This results as follows: $+ 1$ D. S. $\bigcirc - 3$ D. cylinder axis 180°, $- 2$ D. S. $\bigcirc + 3$ D. cylinder axis 90°

In order to verify these transpositions, let us compare them with the diagram on the board.

$+ 1$ D. sphere gives $+ 1$ value in both vertical and horizontal meridians. This is what is desired in the horizontal meridian. When we combine a $- 3$ D. cylinder axis 180°, this horizontal meridian is unchanged because the axis of the cylinder has no refractive value, while the $- 3$ D. against the $+ 1$ D. in the vertical meridian leaves $- 2$ D. value in that meridian, which is just what is desired.

In the second sphero-cylinder, the $- 2$ D. sphere gives $- 2$ value in both meridians. This is correct for the vertical meridian. When we combine with it a $+ 3$ D. cylinder axis 90°, we leave this vertical meridian undisturbed, while the $+ 3$ D. against the $- 2$ D. in the horizontal meridian gives $+ 1$ D. value in this meridian, which is the power desired.

FOGGING METHOD OF EXAMINATION

In any case of refractive error, and especially in astigmatism, it is not well to depend upon a single method of examination, and, therefore, we will proceed to the fogging method. It has received this name because it makes vision indistinct or "foggy." It is intended to simulate cycloplegia, by removing all need for action of the accommodation and placing it as nearly as possible at rest, by having in front of the eye under examination a convex sphere of such strength as to more than overcome any effort of accommodation the eye might otherwise have to make while looking at the letters twenty feet away. The addition of this strong convex lens makes the eye to all intents and purposes, for the time being at least, artificially myopic, and we are then in position to proceed to estimate the amount of error as in any case of regular myopic refraction. You will understand from what I have said that this method is only of value in those cases where a hypermetropic element is suspected, and would not be of service where the eyes are naturally myopic.

We place a $+ 4$ D. sphere in the trial frame and ask the patient to look at the card of radiating lines. In this case, as we have reason to believe that the two principal meridians are vertical and horizontal, we will use the card only on which there are vertical and horizontal lines. The patient says they all looked blurred, but when we question him more closely he says the vertical are darker than the horizontal. From which, of course, we infer that the horizontal lines are least distinct.

Now our plan is to add concave cylinders until we make all the lines equally distinct and black. You will recall that the rule for placing the axis of the cylinder is in the same direction as the indistinct lines. We place a — 1 D. cylinder axis 180° in front of the strong convex sphere, and the patient tells us that the horizontal lines are now a little darker, but not yet equal to the vertical lines. We increase this cylinder .50 D. at a time, and when we reach — 3 D., the vertical and horizontal lines look both alike.

A CASE OF SIMPLE MYOPIA

This cylinder represents the amount of astigmatism, and the case is now reduced to one of simple myopia, which we must correct by the addition of concave spheres.

We now remove the card of radiating lines and replace it with the card of test letters. You will remember there is in the trial frame the original + 4 D. sphere and a — 3 D. cylinder with axis at 180°. Patient says he can scarcely make out the E at the top of card, which is the 200 feet letter. We add a — 50 D. sphere which affords a slight improvement, and we keep on increasing the concave sphere, each change causing a further improvement until we reach — 3 D. sphere, when the vision equals $\frac{20}{20}$.

Now let us see what we have in the trial frame. I will write it on the blackboard so that all of you can follow me.

$$\frac{+ 4 \text{ D. S.} \ \bigcirc - 3 \text{ D. cyl. axis } 180°.}{- 3 \text{ D. S.}}$$
$$+ 1 \text{ D. S.} \ \bigcirc - 3 \text{ D. cyl. axis } 180°.$$

This formula you see is identical with one of the sphero-cylindrical transpositions of the cross-cylinder, and is equivalent to the other sphero-cylinder and to the cross-cylinder.

Inasmuch as it is advisable for you to use a cycloplegic only in very exceptional cases, I would recommend the fogging method to you as the best substitute for developing the latent error in any case of supposed hypermetropic refraction, whether simple or compound. After you have had a little experience with it, you will often be surprised how much of the latent defect you will be able to uncover.

THE STENOPAIC SLIT

In the further examination of this case, we will make use of the stenopaic slit. The patient looks at the test letters (not at the radiating lines as some of you might suppose) and we place the slit in the trial frame at 90°. Patient says the letters are all blurred, and he is unable to name even the larger letters. We begin to rotate the slit with the effect of improving vision and when we get to the horizontal position, patient is able to name the letters on the No. 20 line. This proves that the horizontal meridian is the one of best vision, and it may be either emmetropic or hypermetropic. This point is to be determined by the acceptance or rejection of convex lenses.

We use convex spheres (not cylinders as some of you might suppose) commencing with $+.50$ D., and increasing until rejected. We find that a $+1$ D. is accepted, but a $+1.25$ D. makes the No. 20 line markedly indistinct. This proves then that the horizontal meridian is hypermetropic and to the extent of 1 D.

We now turn the slit around a quarter of a circle to the meridian of poorest vision at 90°. According to our custom, we will commence to test with convex spheres. These are unhesitatingly rejected. We then try concaves, which are quickly accepted. A -1 D. brings the larger letters into view, and we increase .25 D. at a time until we reach -2 D., with which the No. 20 line is legible. This proves that the vertical meridian is myopic to the extent of 2 D.

AXIS AND MERIDIAN

We now have determined the location of the two principal meridians and the refraction of each, and as there is sometimes a confusion in the minds of optical students as to the correct meaning of "axis" and "meridian," I will make a diagram on the board.

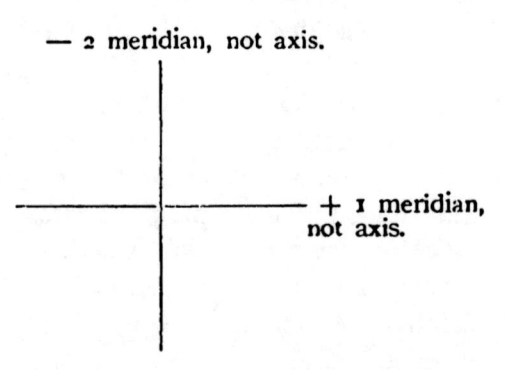

— 2 meridian, not axis.

$+1$ meridian, not axis.

Now you must remember that these are the defective meridians, and that the axis of the correcting cylinders must be placed not in these meridians, but at right angles to them, because the axis of the lens is plane and the refractive power of the lens is at right angles to axis.

The result then of the test by the stenopaic disk is $+ 1$ D. cylinder axis 90° ◯ $- 2$ D. cylinder axis 180°, thus corroborating the other tests.

The stenopaic disk shows a difference of 3 D. between the two meridians, the refraction of the vertical meridian being 2 D. above emmetropia, and the horizontal meridian 1 D. less than normal.

A $- 3$ D. cylinder axis 180° overcorrects the vertical meridian 1 D., and leaves the horizontal meridian unaffected at $+ 1$ D. Hence a $+ 1$ D. sphere combined with this cylinder corrects the total error of refraction.

A $+ 3$ D. cylinder axis 90° overcorrects the horizontal meridian, and makes it myopic to the extent of 2 D. ; and as the vertical meridian is unaffected by this cylinder, both meridians are now equally myopic to the extent of 2 D. Therefore, a $- 2$ D. sphere combined with this convex cylinder will represent the total refractive error.

This case of astigmatism may be classed as regular, mixed, symmetric and with the rule.

As so many beginners in optometry dread meeting a case of mixed astigmatism, I have purposely refrained from using the objective methods, ophthalmoscope, retinoscope and ophthalmometer, and have demonstrated to you how such cases may be worked out by the simpler and more common subjective methods.

Anisometropia

[CLINIC NO. 13]

Mrs. J. C., aged twenty-eight years, complains principally of headache. She tells us she has been wearing glasses for the past ten years, but they fail to afford her much relief. A member of the class has neutralized them for us and tells us they are O. D. $+ .25$ and O. S. $+ .50$, both spheres.

We request her to be seated in the usual testing chair, and placing the opaque disk over the left eye, we ask her what letters she can see on the test card. She replies that she can see only the very large letters, and in response to our request to be more specific, she names the three top letters, E C B, as the only ones she can see. The letters C B you will recognize as constituting the line marked 100 feet, and therefore we will record the vision in this eye as equaling $\frac{20}{100}$.

We transfer the opaque disk from the left eye to the right, and again ask the patient what letters she can see on the card. She commences at the top and names the letters correctly until she reaches the fifth line, the one that is marked to be seen by the normal eye at 40 feet. She hesitates, and as we encourage her to make an effort to name the letters on this line, she replies " P S O E." You will notice that she has omitted the middle letter entirely, and of the four she has named, only one is correct, and that is the last one. We will record her vision as $\frac{20}{40}$, modifying the fraction by four interrogation points placed after it, signifying that four of the letters have been misnamed.

FINDING CAUSE OF IMPAIRED VISION

As there is considerable impairment of vision, our next step will be the use of the pin-hole disk in order to determine if this is a case that can be benefited by glasses. The patient experiences some difficulty in getting the small hole directly over the visual line, but finally she gets it into position, and trying each eye separately we find that vision of right eye is raised from $\frac{20}{100}$ to $\frac{20}{40}$ partly, and left eye from $\frac{20}{40}$ barely to $\frac{20}{30}$ partly. This proves that the sight is susceptible of improvement by glasses; otherwise, it would be useless and a waste of time to try her vision with test lenses.

We will now make an ophthalmoscopic examination to see if that will throw any light on the cause of the impaired vision, which even the pin-hole fails to raise to normal.

There is a dimness and indistinctness about the whole fundus, while the optic disk in each eye is very much paler than normal, which indicates an atrophied condition of the nerve. We would therefore class this case as one of amblyopia in addition to any refractive error that may be present.

Before commencing with any test lenses, we will call the ophthalmometer to our aid. We fix the mires so that they are barely in contact with each other in the primary position, which is horizontal, and as we rotate the instrument the one mire commenees to climb over the other until we reach the vertical meridian, where there is an overlapping of six steps or 6 D. In the left eye there is also a marked overlapping in the vertical meridian, in this eye amounting to three and a half steps, or 3.50 D. This case shows an unusual amount of corneal astigmatism, and I want to give every member of the class an opportunity to see the mires overlapping as they pass from the horizontal to the vertical meridian.

In the examination of vision with the test lenses, we will commence with $+$ 1 sphere and $+$ 1 cylinder, comparing the effect of one with the other. In view of the large amount of astigmatism shown by the ophthalmometer, we are not surprised that patient rejects the sphere and accepts the cylinder. We now increase the lens .50 D. at a time until we reach $+$ 3 D., with which axis at 90° vision equals $\frac{20}{40}$. We hold a $+$ .50 D. and a $+$ 1 D. in front of the cylinder, both of which are rejected. We rotate the cylinder slightly to the right and to the left, with the effect of impairing vision, so that we are justified in concluding that $+$ 3 D. cyl. axis 90° represents the correction of the manifest error.

We cover up the right eye and place $+$ .50 D. sphere before the left : this is rejected and we replace with a $+$ .50 D. cyl. axis 90°, which also is rejected. Bearing in mind the astigmatism which is revealed by the ophthalmometer, and which is shown to be "with the rule," we now try a $-$ .50 D. cyl. with axis at 180°. This is accepted as affording an improvement in the vision, and we cautiously increase it .25 D. at time, until we reach $-$ 1 D. cyl. axis 180°, with which vision equals $\frac{2}{30}$.

We will now use the fogging method to see if the eyes will bear any closer correction. We have already, on a number of occasions, explained this method at some length, and therefore we will not take the time on this occasion to describe every step, but will give you the results as follow:

R. E., + 3 D. sph. ◯ − 3 D. cyl. axis 180°
L. E., + .50 D. sph. ◯ − 1.50 D. cyl. axis 180°

A comparison of this formula with that first obtained shows that the right lens is equivalent in both examinations. In the left eye we have succeeded in having a + .50 D. sphere accepted, with a corresponding increase in the concave cylinder. In order that the two lenses may correspond as nearly as possible, we will transpose and order as follows:

R., + 3 D. cyl. axis 90°
L., − 1 D. sph. ◯ + 1.50 D. cyl. axis 90°

MADDOX ROD TEST

We will complete our examination with the test of the muscular balance, by means of the Maddox rod; which, on account of its poorer vision, is placed over the right eye.

We say to the patient, by way of preparing her for the test, that she sees a red streak running up and down, and ask her on which side of the light it is and how far from the light. She replies that it is on the left side of the light and that it is about an inch away from it. This artificial diplopia we have produced is of the crossed or heteronymous variety, and we know that it indicates an exophoria of small degree. We take a 1° prism and place it over the eye base in, and the red streak is now brought directly over the light, thus showing an exophoria of 1°.

We now turn the Maddox rod half way around, and we say to the patient that she sees the red streak running crosswise, and ask her if it is above or below the light; she replies that it is below the light. We ask her how far below, and she tells us about one inch. This shows a slight degree of right hyperphoria, which is corrected by a prism base down, and when we place a 1° prism in this position it raises the streak so that it runs directly through the light.

I want particularly in this case to call your attention to the difference in the refraction and in the acuteness of vision of the two eyes, which condition is known in optometry as anisometropia.

This is not infrequently met with, and is usually a source of discomfort to the patient, and of annoyance to the optometrist.

When we consider that a very slight change in the curvature of the cornea or in the length of the eye, causes a marked variation in the condition of refraction, it is really to be wondered at that the difference in any pair of eyes is not greater than is usually found; or, in other words, that marked anisometropia is not more common. The statement is made that a variation of $\frac{1}{8}$ mm. in the radius of the cornea or of $\frac{1}{3}$ mm. in the length of the eyeball, causes a change of 1 D. in the refraction of the eye. And while it is seldom possible to find two eyes exactly alike, yet in the majority of cases the difference is so slight as to be detectable only by a careful examination, and therefore small differences may be regarded as more or less physiological and calling for no special treatment. In other words, anisometropia would call for no special consideration at our hands, unless it caused some impairment of binocular vision or some nervous disturbance.

ANISOMETROPIA

The term anisometropia signifies nothing as to the refractive condition of either eye. One may be emmetropic and the other myopic or hypermetropic; both may be hypermetropic or myopic, varying only in degree; one may be hypermetropic and the other myopic; or one eye may be astigmatic, or there may be a greater degree of astigmatism in one eye than in the other.

In the great majority of cases anisometropia is a congenital condition, and is attributable to an unequal development of the two eyes.

Anisometropia may be considered under three different heads:

1. When binocular vision exists.
2. When monocular vison exists, the eyes being used alternately
2. When monocular vision exists, the sight being confined to one eye.

In the consideration of the first class of cases, we must first prove that binocular vision really.is present. This can be easily done by means of a vertical prism or the Maddox rod, while the patient looks at the light across the room.

When a prism is placed before one eye with base up or down, a vertical artifical diplopia should be produced if there is simultaneous vision; but if it is impossible to make patient see two lights with any strength of prism, binocular vision does not exist.

When the Maddox rod is placed before one eye, if patient sees both the streak and the light, binocular vision is present; otherwise not.

In the case we have before us, binocular vision is present as proven by the Maddox-rod test of the muscles, in spite of the difference in the visual acuity of the two eyes, and the character of the astigmatism, one being simple hypermetropic and the other mixed.

In order that binocular vision may be present in any case of anisometropia, there must not be too great a difference in the refraction, and this brings up the question as to the manner in which such simultaneous vision is made possible.

There is a difference of opinion among ophthalmological authorities whether it is possible to equalize the refraction of the two eyes by exercising a greater amount of accommodation in one eye than the other. Some writers argue that such is the case, that one ciliary muscle can act independently of the other, thus correcting each eye by a separate and independent accommodation.

But it is more generally believed that the same amount of nervous impulse goes to each ciliary muscle, and hence that an equal effort of accommodation is made on both sides, with the result that on one retina the image is clear, while on the other it is diffused. Then the act of binocular vision is completed by the brain fusing the blurred image with the distinct one.

It follows then that the eye which possesses the greatest visual acuity, or that requires the least accommodation, will play the chief part in the act of vision, and that the other eye with its blurred image will be a secondary factor, although it must be admitted a not unimportant one.

VARIETIES OF ANISOMETROPIA

The second form of anisometropia is that in which the eyes are used alternately. One eye may be emmetropic or slightly hypermetropic, and is preferred for distant use; the other is myopic, and hence is adapted for near vision.

Such a patient does not possess binocular vision, but at the same time the disadvantages are not all on his side. Both his distant and near vision are fairly good and that without the use of glasses, even though he may have reached the presbyopic age.

The third variety of anisometropia is that in which one eye is permanently excluded, and the other is used constantly both for near and distance. In these cases the inequality between the eyes is very great, the blurred image formed in the poorer eye is ignored, which eye gradually becomes amblyopic and then begins to deviate from its normal position, usually inwards.

Asthenopia may occur in cases of anisometropia, and sometimes it is difficult to determine whether it is due to the refractive error that is always present, or to the inequality between the two eyes; but as a broad general rule, we may state that anisometropic asthenopia will occur only in the first class of cases, and is due to nerve exhaustion in the effort to maintain binocular vision under disadvantageous conditions, as the images in the two eyes differ not only in clearness, but are also unequal in size.

This lady's case comes under this head, the asthenopia showing itself by a constant headache, the difference between the retinal images being such that binocular vision is maintained only by a constant effort and waste of nerve energy.

In the two other forms of anisometropia, where binocular vision is not present, if asthenopia occurs it must be due to the refractive error and not to the anisometropia.

CONSIDERATIONS IN CORRECTION OF ANISOMETROPIA

In the correction of cases of anisometropia, it should always be our aim to give the appropriate lens for each eye. If the difference is slight, such lenses will prove satisfactory; but when the difference is more marked, many persons cannot tolerate such a correction, partly because of the nerve disturbance that is caused when an eye which previously had borne only a subordinate part in the act of vision, is now suddenly called upon to bear an equal burden with its fellow, this being a great change from the condition to which the person had been for so long a time accustomed.

But we must get to as near full correction for each eye as we can, and really our only guide in this matter is the patient's sensations after a careful trial. In this case we feel justified in ordering the full correction in the expectation that after the first two weeks of trial, the glasses will prove satisfactory. It goes without saying, that childhood is the most favorable age for the correction of a case like this. It is most unfortunate that she was not properly fitted ten years ago when she first applied for glasses; the simple spheres she has been wearing have been of little value to her, and they certainly are a reflection on the skill of the person who fitted them. If the proper cylinders had been prescribed at that time, we feel safe in saying that the acuteness of vision to-day would be higher than it is.

In the second class of cases, where one eye is used for distance and the other for reading, as a rule the patient does not suffer from asthenopia, and the attempt to fit him with glasses will be a useless and thankless task. Such a condition of vision calls for but little muscular effort, imposing but slight tax on either the accommodation or convergence.

In the third class of cases, any existing error of refraction or accommodation is to be corrected according to customary methods, while the muscular balance of the two eyes or the refractive condition of the other eye, does not enter into the question.

Convergent Strabismus

[CLINIC NO. 14]

This little girl (Sylvia S.) is five years of age. Her mother tells us the left eye has always turned in, and as you look at this child you have no difficulty in seeing that it is an unmistakable case of convergent strabismus. For several weeks past she has been rubbing spittle on her eyes and complaining to her mother that there is dirt in them.

The frequency with which strabismus occurs in childhood, and the importance of a correct understanding of the conditions involved in order that the case may be properly managed, justifies us in giving time to the careful consideration of this subject.

Strabismus is usually defined as consisting of a deviation of the visual axis of one eye from the correct position of fixation. But this is only a partial definition: In a case of convergent strabismus there is in addition to the abnormal convergence, a defect of the fusion faculty, a suppression of vision in the eye which is not used for fixation, and a condition of amblyopia in the deviating eye, either congenital or acquired as the result of neglect or inefficient treatment. There is also an error of refraction present, usually hypermetropia.

NATURAL DESIRE FOR BINOCULAR VISION

In normal eyes the natural desire for binocular vision causes the visual axes of the two eyes to meet at the object looked at. But if this desire is absent or interfered with, the incentive to perfect accord between the movements of the two eyes is lost, and then any slight cause may disturb the equilibrium of the convergence center and allow the visual axes to assume faulty directions.

In order to see distinctly such a person must fix the object with one or the other eye, and he will naturally choose the eye that has the highest visual acuity or the lowest error of refraction. Both eyes move together, but both are not directed to the same point. He moves his eyes until he gets his best eye in the desired position, and then the other eye will be turning in towards the nose.

Always remember that in convergent strabismus both eyes do not deviate inwards; the good eye assumes a straight position, while the poorer eye shows the convergence of both. The first is known as the "fixing eye," the other as the "squinting or deviating eye." In the majority of cases of convergent strabismus, the separate movements of each eye are perfect; when one eye is covered, the other can turn up, down, in and out to the normal extent.

The natural relation between accommodation and convergence is not disturbed, as proven by the fact that when the fixing eye accommodates strongly for a close object, the deviating eye turns still more inwards, proportionately to the extra effort of accommodation that is put forth.

CONVERGENT STRABISMUS

Convergent strabismus sometimes occurs as occasional and again as alternating strabismus. In the latter case, on account of each eye assuming the burden of vision at times, the vision of both is good, and for this reason such patients do not suffer from neglect.

Many of the students whom I meet have the mistaken idea that in convergent strabismus because the two eyes cannot be directed to the same object, the patient sees everything double. But such is not the case; the image formed in the deviating eye is mentally ignored, all the attention being given to the image formed in the fixing eye. The patient is not conscious of suppressing this impression; it is involuntary and is probably due to a defect in the fusion faculty. Not only is diplopia absent in these cases, but it is usually impossible even to produce an artificial diplopia by means of prisms or different colored glasses before each eye.

In view of the amblyopic condition of the retina of a squinting eye, the question has often been discussed whether the amblyopia with its imperfect power of fixation causes the deviation, or whether the strabismus with its non-participation in the act of vision, causes a deterioration of the retina from disuse.

In the case of convergent strabismus, when the fixing eye is covered the deviating eye becomes straight and assumes the act of vision, its image being no longer suppressed. In some cases its vision is so poor that fingers can scarcely be counted, while in other cases the larger letters on the test card are legible. Sometimes when central vision is almost entirely lost, indirect vision suffers but little impairment, and then by turning the head and eye objects are seen much more clearly.

ERRONEOUS POPULAR BELIEF

There is a widespread belief among the laity as to the spontaneous cure of strabismus, which is probably based upon the fact that when the child attains his full growth, the angle of convergence becomes less without any treatment. But this does not often happen, and when it does the squinting eye has by this time become blind from disuse. I have seen this occur in many, many cases because the parents were advised to wait and see if the child would not outgrow the defect.

The age at which a convergent strabismus first shows itself is an important point. Statistics prove that in seventy-five per cent. of the cases, the deviation appeared before the fifth year, while in a very trifling percentage was its advent delayed until after the sixth year. In more than ten per cent. of the cases the strabismus developed during the first year of life.

In regard to the ætiology of convergent strabismus, many curious suggestions were formerly made by the family and friends. Professionally, the first definite theory attributed it to a shortening of the internal recti muscles, for which the natural cure would be a division of these muscles.

This resulted in a great deal of indiscriminate muscle cutting, the disastrous effects of which were beginning to be recognized when Donders published his great work and gave to the optical world his accommodation theory as to the causation of strabismus.

When a pair of emmetropic eyes are engaged in distant vision, the accommodation and convergence are both at rest. When such eyes are directed to an object close by, they must converge in order that both visual axes may meet at the object, and at the same time they must accommodate in order that the object may be seen clearly.

These two functions, convergence and accommodation, being always performed together, have become "associated," so that it is difficult (and unnatural) to use one without the other.

DONDER'S INVESTIGATION

An error of refraction disturbs this association. A hypermetrope must accommodate even for distance, and still more for near objects. This unnatural and excessive accommodation tends to produce a proportionate abnormal convergence. Donders first recognized this tendency as the cause of convergent strabismus, and he advised correction of the hypermetropia as a cure for the strabismus.

While this was a great advance in optics at that day and led to a rational treatment of this defect, yet in the light of our later knowledge, we cannot unreservedly accept the accommodation theory as the fundamental cause of strabismus.

The vast majority of children are hypermetropic, yet only a very small percentage (perhaps four per cent.) develop strabismus. It has also been proven that the amount of hypermetropia has but little to do with the question as to whether the patient shall or shall not squint in the first instance, but when once strabismus is established, the refractive error becomes an important factor.

We must go farther back than the eye itself in the investigation of this subject, and the most recent views show that the essential cause of strabismus is a defect in the fusion faculty.

THE FUSION FACULTY

At birth we do not find the proper co-ordination of the eyes, but the fusion faculty begins to develop about the sixth month, and is perhaps not complete before the sixth year. At first the instinctive desire for binocular vision will keep the eyes straight, and when the fusion faculty is fully established, it is doubtful if an error of refraction can cause strabismus.

Exceptionally the fusion faculty does not develop until later, or develops very imperfectly, or it may never develop at all; and then it is an easy matter for anything that disturbs the balance of the motor co-ordination, to cause a strabismus.

In these cases then where there is a defect in the fusion faculty, and the eyes are in a state of unstable equilibrium, ready to deviate inwards or outwards on the slightest provocation, an error of refraction such as hypermetropia proves to be an important factor, for the reasons which I have already briefly mentioned to you.

In the great majority of cases of hypermetropia, it cannot be doubted that the fusion sense is unimpaired and hence the eyes maintain their proper positions. But in the minority cases, where the fusion sense is deficient, there is no check to the tendency to deviation and the child develops a convergent strabismus.

At first the strabismus is occasional, manifesting itself only when the child is looking intently at some close object, and disappearing when the accommodation relaxes. It is at this period that the proper convex lenses to correct the hypermetropia are of the greatest value. But if the case is neglected, the excessive convergence becomes permanent and the strabismus is noticeable even when the eyes are completely at rest. At this time convex lenses do not cause an immediate disappearance of the deviation, but if the wearing of glasses is persevered in, there is gradual restoration of the normal position.

EXAMINATION OF THE PATIENT

We will now return to the examination of our little patient. No tests are required in this case to determine the presence of a deviation or its character; a simple inspection shows a convergent strabismus of the left eye. We ask the child with both eyes open to follow the movements of our fingers, and as the two eyes turn equally in all directions, we know the strabismus is concomitant.

The next step would be to estimate the visual acuity, but this is not always possible in the case of young children. We ask the mother if this child knows her letters, and we are informed that she has been attending a kindergarten school and is familiar with some of the letters. After a few patient trials, we find that the visual acuity is about $\frac{20}{60}$, and that she seems to see equally well with either eye.

A subjective test in a child so young as this is scarcely to be depended upon, but we will see what information we can gain from it. After several changes of lenses $+$ 1 D. spheres are accepted, with which vision is raised to $\frac{20}{40}$.

We will now use the ophthalmometer to determine the existence of corneal astigmatism, but the eyes are in such constant motion it is difficult to get definite results. However, we can see that the mires overlap considerably in the vertical meridian, from two to three steps.

The retinoscope, on which we must chiefly depend in estimating the refraction in young children, reveals a hypermetropia of 1.50 D. in the horizontal meridian and 1 D. in the vertical meridian, thus showing a compound hypermetropic astigmatism with the rule.

The ophthalmometer indicates a higher degree of astigmatism than the retinoscope, but as we are unable to verify our findings by a subjective examination, we will use our judgment and order the following as the refractive correction:

+ .50 D. sph. ◯ + .50 D. cyl. axis 90°.

USE OF PRISMS

In strabismus seldom can we measure the amount of deviation by prisms, for the reason that the image formed in one eye is suppressed and the vision is monocular. However, in this case we are more fortunate, for when we place a Maddox rod over the left eye, the child sees the red streak way off to the left, and the light in its proper position. This indicates a high degree of esophoria, and we commence with weak prisms, bases out, which bring the streak and light closer until finally we reach a 20° prism, with which the streak is directly through the light as far as we can depend upon the answers of our little patient.

This would ordinarily call for an increase in the strength of the convex spheres, but as this would blur the vision at first, the child would be inclined to reject the glasses. She cannot understand the value of the glasses, or the reason for which they are given, and if she found she could see better without them, she would probably refuse to wear them.

Therefore we will rely on a prismatic correction of the deviation to a partial extent, and will order a 2° prism base out over each eye to be combined with the sphero-cylinder. We will direct these glasses to be worn constantly, and tell the mother that they should never be removed except for toilet purposes and when the child is in bed.

It is sometimes necessary to prescribe glasses for children even younger than this. Some authorities claim that no infant is too young to wear glasses when required, even if not more than twelve months old, but in my own experience I have never ordered them in a child less than three years old. Of course, such young children often break their glasses, but I have never known the eye to be injured thereby. The lens being confined by the frame does not break into pieces, or if it does they are not dislodged from the frame.

We will direct the mother to have this child's eyes examined once a year, and in this was we can probably keep the strabismus under control and prevent either eye from becoming amblyopic, with the result that the child will attain maturity with a pair of serviceable eyes. Whereas, if a case like this is neglected, the strabismus becomes fixed and the sight of the deviating eye is lost.

Divergent Strabismus

[CLINIC No. 15]

In looking through the cases that have assembled at the clinic to-day, we will pick out this young man, because we notice that he presents a marked divergence of one eye, and therefore he will probably prove to be an interesting case; at any rate, he will serve to complement the little girl we examined at our last meeting, in which there was a marked convergence of the visual axes.

H. K., aged twenty, telegraph operator, complains of twitching of eyes, photophobia, lachrymation and pain after reading.

We see the patient is already wearing glasses, which we neutralize with the following result:

R., + 1 S. ○ + 1.50 cyl. axis 120°; L., + .75 S. As these lenses have been prescribed by a very competent optometrist, they are probably not far from right.

In determining the acuteness of vision of the right eye, in answer to our query the patient names four of the five letters on the number forty line. We therefore record the acuteness of vision as $\frac{20}{40}$?

METHOD OF EXAMINATION

We place a + .50 sphere in front of the eye, and patient is in doubt whether it affords any improvement. We replace it with a + .50 cyl. axis vertical, which is at once preferred as affording the greater acuteness of vision. Leaving this + .50 cylinder in the trial frame, we place in front of it another + .50 cylinder with axis in same position, and alternate with a + .50 S. in order that patient may decide which affords the better vision. The cylinder is unquestionably preferred, and hence we replace the + .50 cylinder with a + 1 cylinder.

We now rotate the cylinder first to the left, which patient says makes vision much worse; then to the right, which is accepted as affording better vision. After a few trials, we find that the proper position for the axis of the cylinder is 110°.

We now have a + 1 cyl. axis 110° over the right eye, which improves vision to $\frac{20}{30}$. We repeat the procedure we went through a moment ago, viz.: placing in front of this lens alternately a +

.50 cyl. and a + .50 sphere. The cylinder is three times accepted and the sphere once, which gives us the following result

+ 1 S. ◯ + 2 cyl. axis 110°, with which vision is $\frac{20}{20}$ clearly.

In comparing this lens with the one he is wearing, you will notice that the cylinder is .50 D. stronger, and that the axis is 10° nearer the vertical. As we have made our examinations carefully, and as the gradual increase in the strength of the lenses was accepted without hesitation, we feel that our stronger cylinder is correct.

In order to determine with accuracy the position of the axis of the cylinder, we will call to our aid the ophthalmometer. This instrument shows an overlapping of 2.50 D. at 110°, which proves that this is the meridian of greatest curvature, while the meridian of least curvature must be at 20°. This verifies the result of the test by the trial case, as to the strength of the cylinder and the proper position of the axis.

As we turn to the examination of the left eye, we are struck with the marked divergence of its visual axis. None of the letters on the test card can be seen, even though we walk the patient up to within two or three feet of the card. The vision of this eye is reduced to the counting of fingers at fifteen inches.

We try the pin-hole disk, and we try a few convex and concave lenses, but the eye does not respond in the slightest particular. This forces us to the conclusion that this eye is amblyopic, that it is beyond the reach of optical assistance and also probably of medical help.

THREE INTERESTING FEATURES

This case combines in itself three interesting features: compound hypermetropic astigmatism, monocular vision, divergent strabismus, to the latter of which we will give a few moments' attention.

In the first place, we wish to determine whether this case is one of concomitant or paralytic strabismus. We hold our finger directly in front of patient and ask him to look intently at it. We now move it slowly first to the right and then to the left, and ask patient to follow with his eyes the movements of the finger. In doing this the squinting eye follows the good eye in all its movements, which shows that there is no paralysis of any of the muscles, and that the strabismus can be classed as concomitant.

At our last meeting we found that convergent strabismus is usually associated with hypermetropia, on account of the close connection between the accommodation and the convergence. And now for the same reason I wish to state that divergent strabismus is usually associated with a myopic condition of the refraction, the history of a typical case being somewhat as follows

As soon as the child begins to attend school, it is found that he has difficulty in seeing the blackboard, perhaps is unable to see a single mark upon it. He is then given a front seat, and the matter is allowed to go at that.

A few years later his myopia has increased so much that he is compelled to hold his book very close, and nothing can be seen upon the blackboard unless he is allowed to approach close to it.

The close position of his book places an unnatural tax upon the convergence. He complains that words run into each other, and he soon tires of close work. About this time he accidentally discovers that by closing one eye he can read comfortably with the other. This encourages him to unconsciously give up the effort of binocular vision by allowing one eye to deviate outwards, thus permitting him to read without effort with the other eye.

At first there is no actual divergence, only a failure of convergence with one eye and monocular vision with the other. This non-use of convergence weakens this function, so that either eye diverges when screened. In these cases there is no diplopia when reading, because the deviating eye takes no part in the act of vision.

In the beginning, when the boy looks up from his book, the divergent eye recovers itself. After a time as the habit becomes confirmed, the eye remains divergent even in distant vision.

TREATMENT OF THE CASE

The treatment of a case of this kind consists in an accurate correction of the myopia. The glasses should be worn constantly, both for near and distant vision. The patient may complain at first that the glasses make print very small, and so they do in contrast with the large diffused image formed in the uncorrected myopic eye; but patients soon become accustomed to them, particularly children.

If the wearing of glasses is postponed until late in life, the accommodation becomes weak from want of use, and in such cases it may be necessary to order an additional pair for reading, 2 D. or

3 D. weaker than the distance glasses. Weaker glasses for reading must also be prescribed in the higher grades of myopia.

In recent cases glasses which afford a normal acuteness of vision usually cause a rapid disappearance of the strabismus, and exceptionally also in cases that have lasted for years. But in many cases where the deviation has existed for some time, which means a prolonged disuse of the convergence, this function has become a negative quantity, and the tendency to divergence cannot be overcome. When the glasses are worn with both eyes open, the fusion faculty prevents any deviation, but if one eye be covered for a moment, it will diverge and remain divergent for a few seconds after the removal of the cover.

This is a picture of the course of a typical case of divergent strabismus caused by myopia, and it is well that you should have a clear understanding of it, but the case under consideration does not come under this head. Here the divergence dates from infancy, and there is a total absence of the fusion sense. The association between convergence and accommodation has been totally destroyed, as shown by a divergent strabismus in connection with a hypermetropic condition of refraction.

In many cases the patient has the power by a considerable effort of convergence to correct the faulty position of the divergent eye, but with a relaxation of such effort the eye at once diverges, and the patient soon finds that this is the most comfortable position. The power of rotation of each eye separately in this young man we find to be normal in every direction, but it is more than likely as time goes on that the power of inward rotation will be very much lessened.

This left eye is practically blind, and it is an interesting fact that when both eyes are blind, they almost invariably diverge. But on the other hand when only one eye is blind, the direction of its deviation will depend to a great extent upon the refraction of the seeing eye. In accordance with this rule, the strabismus in this case ought to be convergent; but the fact that it is divergent instead, makes the case unique and exceptional.

In this connection I may mention an interesting point, and that is in some cases of convergent strabismus in which the deviating eye has become amblyopic, this blind eye in after years may become divergent without any tenotomy having been performed.

MEASURING THE DEGREE OF STRABISMUS

The measurement of the degree of strabismus is a matter concerning which optical students have confused ideas. On asking them this question, many of them answer "by prisms." Now the fact is that very seldom can the amount of strabismus be measured by a prism, for the reason that the vision is monocular. If both eyes participated in the act of vision, diplopia would always be present in strabismus, and then the prism that would fuse the two images would represent the degree of deviation. This would be very simple, but unfortunately in the majority of cases of strabismus, the vision of the squinting eye is suppressed, and then there is no field for a prism.

The simplest way to take the measure of a strabismus is by the angle of deviation of cornea, which is most conveniently done by the perimeter. While this serves all practical purposes, I might say that this measurement determines the apparent, and not the real strabismus. In many cases it is not important to differentiate between the two; especially when the squinting eye is incapable of fixation.

You are all familiar with the appearance of the perimeter. The patient is seated at the instrument, which we adjust so as to bring this left eye accurately in the center of the perimetric arc, while we tell the patient to look at some distant object with the other eye. I now take this lighted candle which I move along the arc until I see the image of the flame as reflected from the cornea in the center of the pupil of the deviating eye. Assuming that the visual line passes directly through the center of the pupil, the number of degrees marked on the arc corresponding to the position of the flame, will represent the measure of the strabismus in degrees. In this case it is 30°.

This young man must take good cure of his right eye, and we will therefore order the sphero-cylindrical correction we found best suited to it. We fear the left eye is and always will remain useless as an organ of vision. However for cosmetic reasons and in order to remove the deformity and allow the eye to act under more normal conditions, I will recommend an operation in this case.

One of two operations may be performed, tenotomy of the external rectus, or advancement of the internal rectus; but in this case I think it will be necessary to do both operations at the same time in order to overcome the excessive divergence and maintain the eye in the proper position.

Diplopia often occurs immediately after a strabismus operation, but is soon succeeded by single binocular vision in a short time if the operation has been done with proper care. But in this case, on account of lack of vision in the squinting eye, we need not fear any such contingency; but at the same time we do not wish to produce a condition of convergent strabismus instead, although the latter is really less disfiguring than the divergent form.

It is impossible to predict what might have been the result if this case had been operated in infancy. It is not beyond the range of possibility that with the straightening of the eye binocular vision might have been produced, and by functional use the acuteness of vision of the eye been raised to normal and maintained there.

Headache in Connection with Myopia and Exophoria

[CLINIC NO. 16]

Miss J. C., aged thirty-three years, complains of headache as soon as she begins to use her eyes for close work.

In addition to the symptoms of which patient complains, it is advisable for you to get a history of the case, more or less complete as its seriousness may demand ; this will often enable you to get a better understanding of the case than would be possible without.

In meeting a patient for the first time, the proper question for you to start with is : "In what way do your eyes trouble you?" The most common answer to this question is "headache."

Then you must make more specific inquiries so as to get more definite knowledge of this one symptom of which so many persons complain.

In what part of the head is it located? Is it in the forehead, in the temples, in the vertex, in the occiput, or does the whole head seem to ache?

At what time of the day does the headache come on—in the morning, in the afternoon or in the evening? Does the patient get up with it, or does it come on later in the day? Is the headache constant, or is it periodic? Does it come on during or after use of the eyes? Does it cease when the eyes are rested, or is it absent altogether when the eyes are not used? Is it better or worse on Sundays as compared with other days? Does the headache come on when shopping, when riding in cars, when attending a public place of amusement, or when in a crowd?

HEADACHES AND EYESTRAIN

No argument is needed to convince you of the frequent dependence of headache upon eyestrain. The cure of headache, oftentimes stubborn and of long standing, by the correction of errors of refraction and of muscular anomalies, is a matter of almost every-day experience with the optometrist in active practice.

These two sources of eyestrain are so often associated that it is really difficult to determine their relative importance, some authorities attaching greatest weight to the strain of the ciliary muscle caused by refractive errors, while others look upon anomalies of the extra-ocular muscles as the greater disturbing factor.

But at the same time we must not allow ourselves to think that every patient who complains of headache is suffering from eyestrain. We must not regard every case of headache as invariably due to a faulty condition of the eyes, nor must we delude ourselves with the thought that glasses can cure all headaches.

With these two thoughts in mind that headaches may be due to eyestrain, and that they may be caused by other conditions of the body entirely separate from the eyes, a careful and thorough examination of the eyes should be made in all cases of continuous or frequent headache, where the cause of the same is not evident, and where the usual medical treatment has failed to afford relief.

In many of these cases the patient is not conscious of any visual defect or asthenopic symptoms, and is apt to assert that his eyes are all right and that there is no use to make an examination of them. In spite of this it has been my frequent experience that an examination has disclosed some refractive or muscular anomaly that has evidently been the cause of the headache, as has been proven by the fact that the correction of the former has been followed by a disappearance of the latter. Experiences like this emphasize the importance of an ocular investigation in every case of headache, even if there are no eye symptoms present.

MANY CAUSES OF HEADACHE

Therefore, it is the province of the optometrist to find out just what part the eyes take in the causation of headache, and not as a matter of routine expect to cure with glasses every case of headache, a considerable number of which may have no direct relation to the eye.

A browache, due to malaria, may be mistaken for ocular headache. Hemicrania, or migraine, or sick-headache, may be an expression of general nervous exhaustion and may bear no relation to the eyes at all. We are keeping well within the bounds of truth when we say that at least one-half of all headaches are due to eyestrain. Gould claims that seventy-five per cent. of all headaches are caused by a faulty condition of the eyes.

The position and character of ocular headaches vary greatly. It may be simply a slight aching or dull pain over the eyes, or at the back of the orbit. It may be an occipital pain, which is suggestive of congestion of the base of the brain. Sometimes the headache is stationary in the forehead, or vertex, or occiput, or it may originate in the brows and pass to the vertex, shooting to the occiput and perhaps even down the spine.

Some authorities assert that in ciliary strain, the pain is generally orbital and frontal; and when the strain is on the extra-ocular muscles, the pain is occipital and spinal; and that temporal headache is due to astigmatism, but I do not attach much importance to these classifications.

The location of the pain in 200 cases of ocular headache has been divided as follows: Eyebrows, 41 per cent.; vertex, 20 per cent.; occipital, 12 per cent.; occipito-frontal, 8 per cent.; temporal, 8 per cent. In one case the headache was general.

It should be remembered that the position may vary with the individual. Some persons when they have a headache, no matter from what cause, always describe it as frontal, others as vertical, and so on. So, that too much dependence must not be placed on the statements of the individual. The commonest form of asthenopic headache is a dull pain over one or both brows, as shown by the analysis of the 200 cases to which I have just referred.

THE PATIENT'S HISTORY

In getting a patient's history, you must not fail to inquire if he or she has ever worn glasses, and if so, how long and with what effect? Make a record also of the character and strength of the lenses, and whether they were prescribed with or without the use of "drops." It is well for us to know whether the eyes have been relieved by the use of the glasses, or whether the same symptoms continue as were present before glasses were worn.

This patient tells us that she has been wearing glasses twelve years, that they were ordered after the use of a mydriatic, and that while they have been of some benefit to her, the headaches have continued and she suffers after use of the eyes for close vision.

We ask one of the young gentlemen to neutralize these lenses, and he tells us they are — 5.50 D. for right eye, and — 2 D. for left eye. We ask him if they were simple spheres, and he replies very emphatically that they are. We return them to him and insist on a more careful examination and neutralization. In looking through the left lens at the straight edge of the test card, he finds there is some displacement when viewed through the optical center of the lens. This proves the presence of a prism, which we find to be 1° placed in frame base in.

In neutralizing lenses with the trial case, a weak prism is very easily overlooked; when a stronger prism is combined, the shape of the edges discloses its presence. When we use the lens measure, we find that one surface is plane and the other curved, and this at once raises the suspicion that the lens is something more than a simple sphere, and we are then led to look for a prismatic element.

MEASURING ERROR OF REFRACTION

We now proceed to measure the error of refraction by trial lenses, in accordance with the methods which I have so often described to you, with the following result:

O. D. — 5.50 S. ◯ — .50 cyl. ax. 90° = $\frac{20}{20}$? ? ?
O. S. — 2 S. = $\frac{20}{20}$? ?

It is impossible with any lens, or any combination of lenses, to raise the vision to $\frac{20}{20}$ full. The lenses mentioned above afford the sharpest acuteness, and you will notice they differ but little in focal strength from the lenses she is at present wearing.

Now, you will remember that the rule in myopia is to give the very weakest lenses with which serviceable vision is obtained, and therefore we will reduce these lenses .50 D., which will make the prescription read as follows:

O. D. — 5 S. ◯ — .50 cyl. axis 90°.
O. S. — 1.50 S.

With these lenses she is able to decipher one or two letters on the No. 20 line, so that they really afford her very fair vision, and, at the same time, they are likely to be somewhat more comfortable than those she has been wearing. Now, these glasses correct the error of refraction and answer for distant vision; shall we instruct her to wear them constantly for all purposes, or will she require some modification for reading? Before deciding this question, let us look into the condition of the muscle balance.

With these lenses in the trial frame, we place a Maddox rod over the right eye. We have a reason for placing it over the right eye, and that is because it is the more defective; the left eye, possessing the better vision, is likely to be the fixing eye.

The rod is placed in a horizontal position, which causes the streak of light to appear vertical. We ask the patient if she sees this red streak running up and down, which side of the light it is, and how far from the light? She replies that she sees the streak and that it appears to be a foot or more to the left of the light. This is a condition of artificial crossed diplopia, which we know must be due to an outward deviation of the eyes, which we call exophoria. We place a prism of 3° over the right eye base in, which she tells us brings the streak closer to the light but still on the left side; prism of 4° brings it still closer; while a prism of 5° base in brings the streak directly over the light.

We turn the rod so as to run vertically, and we say to the patient, in a questioning way, that she sees the streak now running crosswise, and ask her whether it is above or below the light. She replies that it is about two inches below the light, which we know indicates a hyperphoria of this eye. A prism of 1° base down corrects the deviation and brings the streak up on a level with the light.

We repeat these tests at reading distance, using the same Maddox rod and a small point of light, where we find the exophoria has increased to 10°, while the hyperphoria remains the same at 1°. This is only what we expect to find; exophoria is usually greater at reading distance, because as the convergence is called more and more into play its weakness becomes more and more manifest.

This lady is 33 years of age, and her accommodation, which is not very vigorous on account of her myopia, is becoming weakened on account of the physiological changes in the ciliary muscle and crystalline lens, which age brings on; therefore, both her accommodation and convergence need assistance in close vision, which we afford by placing a convex lens of 1 D. over her distance lenses, and combining a prism with the base in.

We will, therefore, order for close use ·

O. D. — 4 D. S. ◯ — .50 D. cyl. axis 90°; prism 3° base in.
O. S. — 1.50 D.

We place the prism over the right eye, because it will thus cause less disturbance of vision than if placed over the left eye, which is the better eye and presumably the fixing eye. In this case such a procedure is better than dividing the prism between the two eyes, and I would advise you always to order the prism over the more defective eye, or if the prism is divided, to place the stronger over the poorest eye.

ANALYSIS OF THE CASE

In analyzing this case, it is reasonable to assume that the headache is due partly to the strain on accommodation and partly to strain of the convergence. In other words, that the asthenopia, of which the headache is the chief symptom, is both accommodative and muscular. This assumption being correct, we will naturally expect the patient to experience relief from the reading glasses, which are so combined as to afford assistance to both functions.

We must impress upon our patient the necessity of changing her glasses whenever she reads or does any close work. This is somewhat troublesome, and as she is able to see fairly well with her distance glasses, she will be tempted many times to read with them, instead of taking the trouble to change. But unless she is willing to use each pair of glasses for its own particular purpose, and change her glasses as often as necessary, she cannot expect to be relieved of her headache.

You will notice that I have corrected only a small part of the exophoria, while the hyperphoria I have ignored for the present. It is well to be slow in the prescription of prismatic lenses; sometimes, even when they are unmistakably indicated, they fail to afford relief. In this case there is such a marked amount of exophoria that we feel justified in correcting a portion of it, and especially as she has worn prisms before. We increase the prism and place it over the right eye instead of the left, as before, in this change expecting to afford greater satisfaction, for reasons already mentioned. If, after wearing the glasses for a sufficient length of time, the patient still complains, I would consider the advisability of increasing the prism or placing another one before the other eye.

For the same reasons I may find it necessary later to consider the advisability of correcting the hyperphoria. When vertical prisms are accepted, they often afford the greatest kind of relief; but more often they are disturbing to vision even though indicated by an existing hyperphoria; therefore, the vertical prism will be held in reserve until we see if we cannot afford relief by the glasses we have just prescribed.

Facultative Hypermetropia

[CLINIC NO. 17]

Miss A. R. B., aged twenty-eight years, stenographer, complains of headache and frequent styes. Says she sees a speck before the left eye like a fly when in a bright light. Also photophobia.

We ask her if she has ever worn glasses: she tells us she got a pair about three years ago, but that she hasn't worn them much, as they proved of no benefit. We neutralize them and find them to be $+ .50$ D.

The first step in the examination of this case (as it should be in the examination of any case of refractive error) is to ascertain the acuteness of vision. We find that she can name every letter on the No. 20 line with each eye separately, although she says it is brighter with the left eye. We record the visual acuity as $\frac{20}{20}$.

We now hand her the reading test card, asking her to look at the small type (.50 D.) at the top of card, with which we find her near point is $6\frac{1}{2}''$ and her far point $20''$.

WHAT THE SYMPTOMS INDICATE

Now, what are we to suspect in this case? Certainly not myopia, on account of the normal acuteness of vision. The headache, the styes, the sensitiveness to light, all point to eyestrain, such as is usually caused by hypermetropia or hypermetropic astigmatism.

When the acuteness of vision is $\frac{20}{20}$ clearly, we are not so likely to find astigmatism, because this error does not allow of perfect definition of objects; therefore, by this process of exclusion we have now narrowed this case down to probable hypermetropia.

How can we determine the presence of hypermetropia, and measure its amount? By two general methods: first, *subjective*, by the use of trial lenses, and dependence upon the answers of the individual under examination. Second, *objective*, by the use of the retinoscope, in which the patient is asked no questions, the examiner depending solely upon his own jndgment.

We will ask one of the members of the class to take the trial case, and ascertain whether the refraction of this case is hypermetropic, and if so the probable amount of manifest hypermetropia.

He commences by placing the opaque disk before one eye, and while this is usually the proper way to examine each eye separately, yet in this case, where both eyes possess normal vision, it is easier and quicker in our preliminary tests, to examine both eyes together.

Our examiner places a pair of $+$ .50 D. spheres before the eyes, and asks patient if she can still read the same line, to which we receive an affirmative answer. Just here I wish to make a remark about the form of the question: do not ask the patient if the lenses make vision better. This would be the proper question to ask in astigmatism or myopia, where the vision is impaired; but in hypermetropia (of the facultative kind), where the vision is not below the standard, the proper way to frame the question is whether the patient is still able to name the letters on the same line.

The lenses are increased .50 D. at a time until $+$ 2.50 D. is reached, when patient says the letters are blurred and she can name only a few of the letters on the No. 20 line, and those only with difficulty. We now drop back to $+$ 2 D. spheres, with which vision is $\frac{20}{20}$, and which represents the amount of the manifest hypermetropia.

The ability to see through convex lenses at a distance proves the presence of hypermetropia, because such lenses make parallel rays of light convergent, and none but a hypermetropic eye can receive convergent rays; and the strongest convex lenses with which the No. 20 line remains legible, is the measure of the manifest error.

THE OBJECTIVE EXAMINATION

So far, so good for the subjective method, and now I will ask another member of the class to take his retinoscope and make an objective examination.

The pupil is illuminated, the mirror is rotated and the shadow is seen to move in the same direction as the movement of the mirror. What does this indicate: either emmetropia or hypermetropia. How are we going to decide which is present: by placing $+$ 1 D. spheres before the eyes to neutralize the distance. If such lenses stop the movement, the case is one of emmetropia; if the movement still continues "with," hypermetropia.

Our examiner tells us the movement is still "with" through the + 1 D. spheres; we direct him to increase the strength of the lenses until the movement is indistinguishable, which he informs us is accomplished by + 3 D. spheres.

After making deduction for the distance at which the mirror is used, the result of the retinoscopic examination is 2 D. of hypermetropia. In this case, therefore, both methods yield the same finding, and the objective verifies the subjective method.

MANIFEST HYPERMETROPIA

This case then is of manifest facultative hypermetropia. Let me stop long enough to explain just what is meant by these terms.

Hypermetropia may be overcome or concealed by the involuntary action of the ciliary muscle, when it is said to be *latent*. When the accommodation is able to relax so that the compensating accommodation or a certain portion of it may be replaced by a convex lens; in other words, when a convex lens is accepted for distance, the hypermetropia is no longer concealed, and is then said to be *manifest*.

This latter is the condition in this young lady's case; the defect is overcome and the vision made good by the action of the accommodation. But as convex lenses are placed before the eyes, the accommodation relaxes and its place is taken by the convex lenses, which in this case are + 2 D., and which represents the amount of the manifest error.

The proportion of hypermetropia that remains latent varies in different individuals and at different ages. In childhood the error is usually all latent, but a gradually-increasing portion becomes manifest with the advance of age and the lessening of accommodative power, until in old age all the latent hypermetropia has become manifest.

Manifest hypermetropia may be either *facultative* or *absolute*. This young lady, whom we know to be hypermetropic at least 2 D., enjoys normal acuteness of vision without any correcting lenses, and she still has normal vision with + 2 D. spheres; in other words, her hypermetropia may be corrected by these lenses or by an equal effort of accommodation. The hypermetropia is manifest because the convex lens is accepted, and we call it facultative because the error can be entirely overcome by the exercise of accommodation.

In thirty-five years from now, when this lady has reached the age of sixty-three years, at which time the amount of accommodation at her disposal will have been reduced to nothing, the hypermetropia will still be manifest, but instead of being facultative, it will now have become absolute. The vision will be considerably impaired, perhaps to $\frac{20}{100}$, because the accommodation is no longer able to neutralize, and then the $+$ 2 D. spheres will be required to overcome the error and raise the vision to normal.

We will now return to the examination of our case, which we roughly determined to be hypermetropic and approximated the amount of defect. We must now test each eye separately, and we can probably get the best results by means of the fogging method.

Covering the left eye with the solid metal disk we place a $+$ 5 D. sphere before the right eye, with which the patient can see only the large letters at the top of the card; that is, with this strong convex lens vision $= \frac{20}{200}$. We gradually reduce with concaves, increasing .50 D. at a time, each change causing an improvement in vision, until we reach $-$ 2.50 D., with which vision has been restored to $\frac{20}{20}$.

Now what have we in the trial frame? $+$ 5 D. and $-$ 2.50 D., which by algebraic addition equals $+$ 2.50 D., which represents the amount of hypermetropia as revealed by the fogging method. We try the other eye in the same way, and with the same result, viz., 2.50 D. of hypermetropia.

TESTING THE MUSCLES

We now proceed to investigate the muscular balance. With these lenses ($+$ 2.50 D.) in the trial frame, we place a Maddox rod over the left eye in a horizontal position. We direct the patient to look at the point of light across the room, saying to her that she will see a red vertical streak and asking her on which side of the light it is and how far from the light. She replies that it is on the right side of the light and about three inches away. The images of the eye being crossed indicates exophoria, which is corrected by a prism base in, and we find that a 2° prism brings the red streak back to the light.

You will probably recall from your studies of muscular imbalances that esophoria is usually associated with hypermetropia, and exophoria with myopia. In this case then we have an anomalous condition, exophoria with hypermetropia, showing a disturbance of the normal relation that should exist between the functions of accommodation and convergence.

We will also make a test of the insufficiency of convergence at the reading distance, where we find 8° of exophoria, the so-called "exophoria in accommodation." This is not uncommon, in fact, exophoria is always greater at the reading point than at a distance, because the closer the object is held the greater the amount of convergence that is necessary, and hence any insufficiency of this function becomes the more manifest.

We will make a further investigation of the muscles by means of the duction tests for determining the relative power of convergence and divergence.

The patient is directed to keep her eyes fixed on the light, while we place prisms bases in, gradually increasing their strength until diplopia is caused, when we return to the strongest prism base in with which single vision can be maintained, which will represent the power of abduction or divergence, which in this case we find to be 6°. This is about normal.

Then we place gradually increasing prisms before the eyes bases out until diplopia is produced, which we encourage the patient to overcome by looking inwards towards the nose, by "trying to look crossed-eyed," as we express it, and we find that 8° bases out are the strongest prisms with which single vision can be maintained, and this therefore is the measure of the adduction or convergence. You will recall that the normal power of adduction is from 20° to 30°; and therefore in this case it is insufficient. This means an absolute weakness of the internal recti muscles, which accounts for the exophoria.

THE GLASSES TO PRESCRIBE

Now, then, what glasses shall we order for the correction of this case? I have always advised you to be slow in the prescribing of prisms, and if possible to influence the muscular imbalance through the refraction, but such advise will not hold good in a case like this. Convex lenses tend to increase exophoria, and if we prescribe full correction for the hypermetropia, we will certainly aggravate the muscular insufficieny.

But we must prescribe convex lenses for the hypermetropia, not allowing them to be too strong however. I think + 1.50 D. would be about right; this is making some concession to the exophoria, and at the same time corrects the major part of the error of refraction.

Shall we dismiss the case with the prescription for these convex lenses? No, we think that prisms should be combined with them to assist the convergence, and we will order as follows

$$\left.\begin{array}{c} \text{O. D.} \\ \text{O. S.} \end{array}\right\} + 1.50 \text{ D. S.} \bigcirc \text{prism } 1° \text{ base in.}$$

We will direct these glasses to be worn constantly, and we prescribe them with the greatest confidence that they will afford relief.

This is a clear-cut case ; there is no doubt as to the condition of the refraction and the muscle balance. The patient has answered the questions without hesitation and without contradicting herself, and the presence of hypermetropia and exophoria is fully established.

The patient asks us how long these glasses will last her ; this is a question no one can answer with certainty. We would advise a re-examination in two years, and sooner if the symptoms seem to call for it.

Hyperphoria

[CLINIC No. 18]

A. H. J., thirty-seven years of age, a machinist by occupation, complains of headaches at times and a soreness and aching in eyeballs. In reading he must make a conscious effort to adjust his eyes for the print. Has felt for some time that his eyes needed attention, but has simply neglected the matter. Has never had his eyes examined, and has never worn glasses.

We find the acuteness of vision a little better than normal, as he is able to name about half the letters on the No. 15 line. In ascertaining his range of accommodation, we find that by an effort, he is able to read the smallest print as close as 8 inches and as far away as 28 inches.

These findings practically exclude myopia and astigmatism, and narrow the refraction down to a choice between emmetropia and hypermetropia. How are we going to determine which of these two conditions is present? The most practical and easily applied methods at our command are the trial case and the retinoscope.

In using the test lenses the diagnosis of hypermetropia depends, of course, upon the acceptance of convex lenses for distant vision, while the rejection of the same would indicate emmetropia.

METHOD OF EXAMINATION

We place a pair of $+$ 1 D. lenses before his eyes, knowing that if vision equals $\frac{20}{40}$ or better, with such lenses, hypermetropia is present. In answer to our question as to whether he can see the same line, patient replies in the affirmative. Mark you, we do not ask if the glasses make vision better ; this is not the proper form of question when the vision is already normal. In myopia, and even in astigmatism, where the vision is greatly impaired, and where our effort is exerted to raise it to normal, we very properly ask patient if such and such lens improves vision, and whether one is better than the other. But in hypermetropia, when the accommodation is able to maintain the vision at the normal standard, there is no question of improving vision by convex lenses, but rather the ability to see through convex lenses as well as without them.

Patient tell us he is able to see through these lenses very clearly, naming the same letters as before; this proves hypermetropia, and in order to determine its amount, we increase the lenses .50 D. at a time until + 2.50 D. is reached, which produces a notable blurring of the letters. We therefore estimate the probable amount of hypermetropia at 2 D.

Having thus determined the condition of refraction with both eyes together, we now proceed more carefully to measure the amount of defect, by testing each eye separately.

As is our usual custom in hypermetropia, we use the fogging system. We place + 5 D. lens in front of the right eye, with which even the No. 200 letter is illegible. We partially neutralize with concave spheres, commencing with — .50 D. and increasing .50 D. at a time, each change of lenses producing a still further improvement of vision. In this method of testing, where the vision is so greatly fogged by the strong convex lenses, it is perfectly proper to ask as we place the concave lenses if they improve vision, as our effort now is to raise the vision to normal in spite of the convex lenses.

When we reach — 2 D. the vision has become normal, some of the letters in the No. 15 line being again legible. The result of — 2 D. placed over + 5 D. is + 3 D., which represents the amount of hypermetropia we have uncovered. We repeat the same test with the left eye, where we find 2.50 D. of hypermetropia.

RETINOSCOPIC TEST

We will now turn to our test by the retinoscope. The direction of movement of the shadow in both eyes is unquestionably "with." This may mean either emmetropia or hypermetropia. In order to determine between these, we place + 1 lenses before the eyes; if these neutralize the movement, emmetropia is present. But if the movement is still "with," the refraction is hypermetropic.

In this case we find the movement is still in the same direction, thus classing the refraction as hypermetropic. We now measure each eye separately, increasing the strength of the convex lens as long as the movement continues "with," and we find that a + 4 D. neutralizes the movement in the right eye, and a + 3.50 D. in the left eye. We subtract 1 D. to allow for the distance at which the test is made, the result being exactly the same as that found by the trial case.

This is quite a marked amount of defect, and it imposes a tremendous tax upon the accommodation, which at this age is scarcely able to bear it. A hypermetropia of this amount at this age would be sufficient to account for the headaches, the aching in eyeballs and the effort required to adjust the eyes for reading.

But we must not be content to stop here, we have finished only half our examination, and as conscientious optometrists we are bound to make the examination of each case thorough and complete.

We therefore pass on to an investigation of the muscular equilibrium, for which we depend on the use of the Maddox rod. This is placed before the left eye in a horizontal position, causing this eye to see a vertical streak of light. In answer to our question as to which side of the light the streak appears and how far from the light, the patient tells us about a foot or more to the left. This being on the same side as the eye over which the rod is placed, indicates esophoria, and is correctible by prisms, bases out, the amount of prism required being the measure of the esophoria, which in this case we find to be 8°; in other words, a prism of this strength is required to bring the streak up to the light. While we would expect to find some esophoria in a case of hypermetropia like this, we are scarcely prepared to find so much.

We now turn the Maddox rod around to the vertical position, when the image formed in this left eye will be a horizontal streak of light. We ask the patient if this streak is above, below or through the light. He answers that it is away below, at least six inches. This indicates a hyperphoria of this eye and is corrected by a prism, base down. The degree of prism required to bring the streak up to the flame will be the measure of the hyperphoria. We commence with a 2° prism, which brings the streak somewhat closer, but still considerably below. We increase to 3°, then to 4°, and then to 5°, when patient tells us the streak is now through the light. We make the entry in our record book L. H. (the abbreviation for left hyperphoria), 5°.

INTERESTING FEATURE OF THE CASE

This throws a new light upon the case, and instead of being one of hypermetropia with its accompanying esophoria, we are inclined to regard the hyperphoria as the essential feature of the case.

I have seen the statement made by an optical writer that in high convergent strabismus there is almost always in addition an upward deviation of the squinting eye. I have not been able to verify this statement in my own experience, but it is well for you to bear in mind the possibility of such connection. This may serve to explain the hyperphoria in this case occurring in connection with a marked degree of esophoria.

A hyperphoria of 1° is capable of giving rise to asthenopic symptoms; much more so than an equal amount of esophoria or exophoria, and therefore this form of deviation calls for our careful consideration.

The average strength of the vertical muscles is scarcely more than 2°; a hyperphoria of 1° is one-half the total strength of these muscles, and hence it can be easily understood why a hyperphoria of this comparatively small amount is such a disturbing factor.

The external and internal recti, on the other hand, are much stronger, and besides the action of these muscles is influenced by the effort of accommodation. This explains why a higher degree of exophoria or esophoria produces much less marked symptoms of asthenopia. And then again, we can at least partially correct the former with concave lenses and the latter with convex lenses, through their action on the accommodation, but we have no such means of influencing the hyperphoria.

When we come to inquire into the etiology of hyperphoria, the only explanation that can be given for its occurrence is that the balance of power between the superior and inferior recti muscles of one eye differs slightly from that of the other.

Hyperphoria may be latent just as hypermetropia is latent. It is that portion of the anomaly which the observer fails to find. What is latent at one time to one man, is manifest to another under other conditions. A relatively greater amount of defect may be latent in the superior and inferior recti than in the external and internal recti, for the reason that the latter in the performance of their function are alternately converging and diverging the optic axes, while the superior and inferior recti are never called upon to produce any change in the relative position of the two optic axes. Whatever position they assume, they must not deviate from the same horizontal plane; consequently, the impulses to the vertical muscles become more fixed, and for this reason defects in these muscles are less likely to manifest themselves.

TREATMENT OF HYPERPHORIA

The treatment of hyperphoria may by optical or surgical. In the higher degrees of hyperphoria, an operation may be advisable, if the symptoms are of sufficient gravity to justify operative interference, either tenotomy of the superior rectus of the hyperphoric eye or advancement of the inferior rectus. But even in these cases we would advise a trial of prisms first, in order to note their effect. If prisms do not afford a fair measure of relief, there is always room for doubt whether an operation will be any more successful. For it must be remembered, that the existence of a hyperphoria cannot by any means be accepted as positive proof, that this is the cause of the headache, or asthenopia or other reflex nervous disturbances of which the patient may complain.

In the optical treatment of hyperphoria, some interesting questions arise in the prescriptions of the correcting prisms. In cases of 1° of hyperphoria a single prisms suffices, and the question occurs over which eye it shall be placed. If the vision of one eye is decidedly poorer than the other, we preferably place the prism over the more defective eye. In cases where the vision of the two eyes is about equal, it has been customary to place the prism over the left eye.

Now, my experience has been that a prism, base down, is not so comfortable as one base up. In other words, the raising of objects by the base-down prism is apt to cause more disturbance of vision than the lowering effect of the base-up prism, probably because the lines of vision are more often and more naturally below the level of the eyes than above.

For this reason, unless there are indications to the contrary, I think it is better to place the prism base up over the cataphoric eye rather than base down over the hyperphoric eye. For instance, in a case of right hyperphoria of 1°, instead of ordering prism base down over right eye and raising the vision of this eye to the level of the other, I would order prism base up over left eye and thus lower the vision of this eye to the level of the right.

There are other indications for varying the position of the prisms, as, for instance, the following: A case of left hyperphoria, of which 1° is shown, when the Maddox rod is over the right eye, and 1½° when over the left eye. In this case the right eye is probably the fixing eye, and a prism of 1°, base up, over this eye would be likely to cause more disturbance of vision (at least, when first worn) than one of base down over left eye, for obvious reasons.

NECESSITY OF PRISMS

The necessity of prisms for the correction of hyperphoria is often easy of demonstration. While the patient looks at the test card through his refractive correction, a vertical prism in its proper position is placed over one eye and then in a moment reversed, when the patient can quickly decide which position is comfortable and which disturbing to vision. In the absence of such indications, that is, if patient is unable to decide which position is comfortable, a prism would be of doubtful value.

When the hyperphoria is 2° or more, it is customary to divide the prism between the two eyes, base down before one eye and base up before the other. This applies to the case under consideration, and we will order 2° base up right eye and 2° base down left eye, thus depressing the image of one eye and elevating that of the other, in this way restoring the visual lines of the two eyes to the same level.

We are now ready to order the glasses for our patient. We must, of course, prescribe for the hypermetropia, but as he has never worn glasses we cannot make them too strong. Nor do I think it well to order a full correction for the hyperphoria. The esophoria for the present can be safely ignored. With these considerations our prescription will read ·

O. D., + 2 D. sph. ◯ prism 2° base up,
O. S., + 1.75 D. sph. ◯ prism 2° base down,

which we will direct to be set in spectacle frames and worn constantly.

A Case of Astigmatism, Illustrating the Value of the Ophthalmometer

[CLINIC No. 19]

Mabel M., aged sixteen years, in good general health, has a.ways had trouble with her eyes. Complains of headache, aching of eyeballs and dimness of vision.

This young lady has been sent to us by one of our former students, with the statement that he recognizes it as a case of some difficulty, that it seems to him to be spasm of accommodation, and that he does not wish to assume the responsibility of prescribing glasses, as he confesses he does not understand the case.

He further tells us that the girl comes to him wearing — .75. D. cylinders, ordered by some other optician, which glasses have proved of no benefit.

USE OF TEST LENSES

In using test lenses, we always commence with convexes; cylinders if astigmatism is present, spheres if astigmatism is absent. In the first examination of a case, we do not know if the refraction is myopic or hypermetropic, but as the majority of cases are the latter, convex lenses are usually accepted; but if rejected, no harm is done. Whereas, if convex lenses are tried first, they are very often accepted even in the presence of hypermetropia, especially if the error is not of high degree.

An eye instinctively makes an effort to overcome concave lenses when placed in front of it, the ciliary muscle is thus called into action and may result in spasm of accommodation, which is essentially a condition of false or artificial myopia, and for this reason concave lenses are accepted and cause an improvement in vision. Therefore, the fact that a patient accepts concave lenses is not proof that he has myopia, because such lenses really produce a condition that favors their acceptance.

The rule to always try convex lenses first and the reasons for it, as I have just stated, are doubtless well known to you all, and yet it seems necessary to mention and emphasize them occasionally, lest you become careless about their observance.

We find the acuteness of vision in this case is 20/60 with each eye separately, and that the range of accommodation with small print is from 3½″ to 9″

Now here we have a case in which the distant vision is impaired and the reading is held close to the eyes, the far point being located at 9″. This condition of distant and near vision makes us think of myopia, because we know that the essential features of this defect are an impairment of distant vision and a holding of the book close to the eyes.

This might lead us to commence our test with concave lenses, presuming the case to be one of myopia; but I admonish you all to resist any such temptation under like conditions.

Let me call your attention to a discrepancy between the amount of possible myopia indicated by the far point of reading vision. With an acuteness of vision of 20/60, there could not be a myopia of more than 1 D. or 1.50 D., whereas a far point of nine inches indicates a myopia of 4.50 D. This discrepancy should at once raise your suspicion that you have not a case of simple myopia to deal with, and if there were no other reason should be sufficient to prevent you from beginning your test with concave lenses.

We now take a + .50 D. sphere and a + .50 D. cylinder, and place first one and then the other over the right eye, rotating the cylinder through the several meridians. The patient tells us that neither lens or any position of the cylinder makes any difference in vision either for the better or worse.

This is not encouraging, but still we must not try concave lenses. What method of examination can we make use of to throw some light on this case, the refraction of which is so much in doubt? We will use the ophthalmometer. In my private office it has become my routine habit to use this instrument after ascertaining the acuteness of vision and the range of accommodation, and before commencing with the trial lenses or the retinoscope or ophthalmoscope.

USE OF THE OPHTHALMOMETER

Perhaps you may be inclined to ask the reason why I use the ophthalmometer so early in the examination of every case, and my reply is that I wish to determine the presence or absence of astigmatism, and when this is done, much time is saved and the case is simplified.

Ordinary cases of hypermetropia and myopia are as a rule easily detected and corrected; but even here certain definite methods of examination should be followed. Perhaps some of you may think that the ophthalmometer is not of any value in case of simple hypermetropia and myopia, but I hasten to assure you that it is invaluable in every case of refractive error that passes through your hands. In case of hypermetropia and myopia, it practically eliminates the question of (corneal) astigmatism. This is a most important matter, for when we know that astigmatism is absent, our trial case examination is more easily and quickly made. Therefore, even in simple axial errors, the ophthalmometer is of great value in furnishing negative information.

Astigmatism is one of the most common refractive errors, and is the cause of much trouble both to patient and prescriber, especially when occurring in the compound and mixed forms. In the correction of refractive errors, astigmatism stands first in importance. Therefore, if we have any ready method by which we can detect its presence early in the examination, it is well for us to take advantage of it, and such a method we have in the ophthalmometer.

We place our young patient seated before the instrument, with her chin upon the chin-rest and her forehead pressed against the head-rest, so as to keep her head fixed and immovable. Both eyes wide open, and the black cover in front of left eye, while we ask patient with right eye to look directly into the telescope. The two eyes should be on the same level, as otherwise the least rotation of the head will change the axis 5° or 10° from its proper position, thus leading the examiner astray and perhaps throwing an undeserved blame upon the instrument.

We now look through the telescope, moving it from one side to the other and up and down, until we have the cornea in the center of the field. Then we focus by moving the instrument closer to and farther from us until the mires are at their most distinct point.

PRIMARY AND SECONDARY POSITIONS

Now we obtain the "primary position," which is in the meridian of least curvature, and is that point where the transverse guide lines are coincident and form one continuous straight line, thus indicating the position of one of the principal meridians.

We rotate the telescope so that the mires are in the horizontal position, where in the majority of cases the guide lines are continuous.

But in this case the lines are broken, and we slowly rotate the instrument until the lines become continuous. This we soon find to be at 20°, and this then is the primary position in this case, and one of the principal meridians. It is now necessary to note the position of the mires; they must not overlap nor must they be separated, and hence we move them slightly so that they are barely in contact.

We are now ready to obtain the "secondary position," which is exactly at right angles to this and is the other principal meridian, or the meridian of greatest curvature.

We rotate the telescope slowly, and we see the one mire creeping over the other more and more, and when we reach the one hundred and tenth meridian, the guide lines have again become continuous, and the mires have overlapped to the extent of five steps or 5 D.

We can now place the blind in front of the right eye, and turn the telescope so as to cover the left eye. Going through the same steps as with the other eye, we find the primary position to be at 160°, and the secondary position at 70°, where the mires overlap to the extent of five steps or 5 D.

Now what have we found: we have ascertained the presence of astigmatism, the amount of astigmatism, the location of the meridians of least and greatest curvature, and the position of the axis of the correcting cylinder.

As to the amount of astigmatism as indicated by the ophthalmometer and the strength of the cylinder accepted by the patient, some little allowance must be made. In astigmatism "with the rule," the ophthalmometer's indication is from .50 D. to 1 D. greater than it should be, whereas in astigmatism "against the rule," the instrument's readings are .50 D. to 1 D. less than they should be.

SYMMETRIC ASTIGMATISM WITH THE RULE

This case then is one of symmetric astigmatism with the rule. We call it symmetric, because the axis of right cylinder is just as far to the right of the median line as the axis of the left cylinder is to the left; or in other words, the combined value of the two meridians equals 180°. We call it "with the rule," because the excess of curvature is vertically.

Perhaps the question may occur to some of you as to why the mires in this case have overlapped as we turn the telescope from the primary to the secondary position. It is simply because of the increased curvature in the vertical meridian, or rather of the lessened radius of the curvature, which crowds the doubled images of the mires into less space, which can be accomplished only by overlapping at their inner edges.

Now, with this definite information gained from the ophthalmometer, the trial case examination is much more simplified. We no longer have to try one lens after another to the weariness of the patient and the tiring of her eyes, but we have quickly determined by an objective method the condition of refraction, which we can verify or disprove bv the subjective method without the trying on of innumerable lenses.

I place a + 4 D. cylinder in the trial frame over the right eye with the axis at 110°. This at once causes a marked improvement in vision, and enables the patient to read the No. 30 line. I rotate the cylinder first one way and then the other, with the effect of producing great impairment of vision. Our patient without any hesitation decides on 110° as the best position for the axis of the cylinder.

I hold a + .50 D. and a + 1 D. sphere in front of this cylinder, both of which spheres are promptly rejected. I place a + .50 D. cylinder over it with axis in same position, which our patient hardly knows whether to accept or reject, but finally decides it is better without.

I then try a — .50 D. cylinder with axis at right angles, which is at once accepted and raises vision to $\frac{20}{20}$. I rotate the concave cylinder so that its axis coincides with that of the convex cylinder, and this position is not as good as when their axes are at right angles.

I make my test of the left eye in the same way, using the strength of cylinder and the position of axis indicated by the ophthalmometer, and after one or two trials, I quickly find the correcting lens, and am now in position to prescribe as follows :

O. D., .50 S. ◯ + 4.50 D. cyl. axis 110°
O. S., + 4.50 D. cyl. axis 70°

which lenses as a matter of course must be worn constantly.

This is really a difficult case, especially so for a man who depends upon his trial lenses, for the reason that the weaker lenses with which it is customary to commence our examination, are so far from the proper correction that they produce but little effect on vision, and the examiner gropes around in the dark trying aimlessly spheres and cylinders, convex and concave, with no result at all or most likely an incorrect prescription.

IMPORTANCE OF THE OPHTHALMOMETER

How different in this case; before we try any test lenses at all, we use the ophthalmometer, and this at once indicates the character and extent of the error, and points out the lens to be used and how it should be placed before the eye.

Perhaps every case may not be so clear and positive as this one, but let me give you a few general directions for putting glasses in the trial frame after the ophthalmometer indicates astigmatism with the rule as in this case.

1st. Try convex cylinders alone, placing the axis in the position indicated by the instrument and about .50 D. weaker than the overlapping calls for. Increase strength of cylinder as long as patient accepts, or decrease slightly if necessary. Rotate cylinder slowly both to right and left, in order to be sure that you have the correct position for the axis.

2d. After having determined the convex cylinder that affords best vision, try convex spheres in front of it, making the effort to have them accepted.

3d. In case the vision is not made perfect by the convex cylinder or sphero-cylinder, try concave cylindrical lenses with axis at right angles, cammencing with — .25 D. and gradually and slowly increasing same until normal vision is reached.

This would be particularly indicated if in a case like this one under consideration, where the mires overlapped 5 D., and the patient would accept only + 2 D. cylinder axis 90° with improvement in vision, all additional convex cylinders and convex spheres being rejected, we would at once suspect mixed astigmatism and proceed to try concave cylinders with axis at right angles to the convex cylinders.

4th. If convex cylinders are positively rejected, then concave cylinders with axis in same meridian as the primary position.

5th. If such concave cylinders fail to raise acuteness of vision to normal, then the cautious addition of concave spheres.

I would suggest to you the following routine of examination

    1st.   Ascertain acuteness of vision.
    2d.   Measure amplitude of accommodation.
    3d.   Use of ophthalmometer.
    4th.   Trial case examination.
    5th.   Muscle tests.
    6th.   Ophthalmoscope.
    7th.   Retinoscope.

If I have impressed upon you the great assistance offered by the ophthalmometer in the early stages of an examination, as demonstrated by this case, I have succeeded in my purpose to-day; and as our time is exhausted, I will not detain you further to complete the remaining steps of the examination.

A Case of Hypermetropia, Illustrating the Fogging Method

[CLINIC No. 20]

J. A. H., twenty-five years of age, machinist by occupation, complains principally of headaches.

We find the acuteness of vision even better than normal, as the patient can name at least half the letters on the No. 15 line. This practically excludes astigmatism; and, of course, myopia is not to be thought of.

But in accordance with the routine method of examination which I advised at our last clinic, we will first make use of the ophthalmometer, which shows a slight overlapping in the vertical meridian. As this is the normal condition of the cornea, that is, as the curvature of the vertical meridian is usually a little shorter than the horizontal, we may assume in this case, at least as far as the cornea indicates, that astigmatism is absent.

We now ask our patient to return to the chair facing the test cards in order that we may make a trial case examination.

We have excluded myopia and astigmatism, and therefore we have no use in this case for concave lenses or cylinders. The visual acuity not being below the standard, the refraction must be either emmetropic or hypermetropic, and in order to determine this question quickly, we place a pair of + 1 D. spheres before the eyes, asking the patient if he can still see the same line. He replies that he can and names the same letters he saw before. Therefore this case is proven to be one of hypermetropia by the acceptance of these convex lenses; and it only remains for us to measure the amount and to test each eye separately.

THE FOGGING SYSTEM

As several members of the class have asked me about the "fogging system," and expressed a desire for instruction in its use, I will embrace this opportunity to demonstrate it.

And, first, what is the fogging system? Many students seem to have a hazy idea of it, while other practitioners of optometry who perhaps have some knowledge of the method, do not appreciate its full value ; so that between ignorance and undervaluation of the fogging system, it is probably not used nearly as much as it should be.

The fogging system is a subjective method of determining the full amount of hypermetropia in an eye under examination, by an endeavor to produce a relaxation of the accommodation, which is concealing a certain amount of latent defect.

Latent hypermetropia is the bugbear of the refractionist, and in the endeavor to uncover it, the optometrist makes use of the fogging system, while the oculist falls back on a cycloplegic. The advantages and disadvantages of "drops" have been bitterly argued by medical and non-medical refractionists, and we will not attempt to decide which side is right, but I want to thoroughly instruct you in this method, which in so many cases detects the latent defect and renders the use of atropine but seldom necessary.

In a hypermetropic eye as soon as opened, the accommodation instinctively comes into action in order to afford perfect vision, which is accomplished by an increase in the refractive power of the eye. This being continued year in and year out becomes a fixed habit which is not easily abandoned.

We wish to supplant the insufficient refraction, or in other words correct the hypermetropia, by means of convex lenses. In order to allow us to do this, the ciliary muscle must retire from the field of action ; but on account of its fixed habit of contraction, this is impossible for it to do at once. Therefore the convex lens is rejected, because it in connection with the active accommodation, produces an excess of refractive power and a consequent blurring of vision.

We overwhelm and disconcert the accommodation by means of a strong convex lens. The strength of the lens needed will vary in different cases, but it should be strong enough to make the large letter E (marked No. 200) barely discernible, so that the patient is in doubt whether it is an E or an F or a B or an H.

The optometrist should then busy himself with some little matter at his desk, while the patient is instructed to quietly look at the letter for a few minutes in an effort to determine exactly what it is. In most cases the letter gradually becomes entirely legible, and he can perhaps even make a guess at the two letters on the 100 line. This proves that there has been some relaxation of the accommodation.

EFFECT OF A STRONG CONVEX LENS

You will understand that when a strong convex lens is placed before the eye, its refractive power is greatly increased, the rays of light are brought to a focus in front of the retina and a condition of artificial myopia produced. Any contraction of the accommodation under such conditions only makes vision worse; while any relaxation would afford an improvement in vision. This furnishes an incentive to relaxation, as the natural instinct is for clear vision, and in this way with a little patience, some lessening of the ciliary contraction may be expected to occur. The strength of the convex lens may now be gradually decreased until the patient's vision is raised to $\frac{20}{20}$. By this method we have tempted the accommodation to relax, or we might say we have caught it off its guard and captured a large part, if not all the hypermetropia.

Now a little attention as to the details of the procedure. In the first place, instead of an opaque disk before the eye not under examination, place an exceedingly strong convex lens, which while affording no real vision of the letters, at least allows the blurred effect of the light and card being seen at the usual distance of twenty feet, thus favoring a relaxation of the accommodation much more than an opaque disk would do.

In the second place, do not reduce the convexity before the eye by substituting weaker convex lenses one after the other. Perhaps you have started with $+ 5$ D., and after allowing a few minutes for the eye to adapt itself, you have replaced it with a $+ 4.50$ D., which again you have replaced with a $+ 4$ D., and so on until you have reached a lens that permits of the normal visual acuity.

If this has been your idea of the proper procedure in the fogging system, you have been on the wrong track, as this is not the correct manner of reducing the strong convex lenses. Let us consider this matter in detail.

You start with a $+ 5$ D. lens with which you expect to coax a relaxation of the ciliary muscle. Then you remove it and replace it with a $+ 4.50$ D. Now what have you done? As soon as the $+ 5$ D. is removed, all restraining influence upon the accommodation is taken away and is stimulated into action again to focus the letters upon the retina. Then you put on the + 4.50 D. for the purpose of restraining the accommodation; then you remove it and again give the accommodation full swing. Thus you alternate from + 5 D. to nothing, then to + 4.50 D., then to nothing, then to + 4 D., again to nothing, and so on. Can you expect to restrain and relax the accommodation in this way? Are you not rather stimulating it into activity? You aggravate the very condition you are endeavoring to modify by such a method of procedure; and you get no better results, if as good, as the usual method of commencing with the weak convexes and gradually increasing their strength.

THE PROPER PROCEDURE

Perhaps you are commencing to ask how you shall proceed, and I am glad to be able to instruct you in the proper way. After you have placed the + 5 D. before the eye, leave it there; do not remove it for any purpose, but instead get the effect of + 4.50 D. by placing — .50 D. in front of it. Then increase this latter lens to — 1 D., then to — 1.50 D. and so on as necessary. In this way the restraining influence of the original convex lens is constantly before the eye, and no opportunity is afforded for stimulation of the accommodation, but on the other hand every inducement for it to become passive.

The concave lens is increased until normal vision is afforded, and the difference between the two lenses will represent the amount of error that has been uncovered. You can readily see that this is the only proper way to employ the fogging system in order to gain its advantages. In fact, it is not fair to apply this term to the method first described. Therefore, if some of you have failed to get satisfactory results from the fogging system, you should not condemn the method itself, but rather carefully scrutinize your manner of conducting it to see if you have been sufficiently careful, accurate and painstaking.

We will now make use of the fogging method in the case before us. We want to reduce his vision to $\frac{20}{200}$ scant. We place this + 6 D. before the right eye and a + 10 D. before the left. These lenses fog his vision to such an extent that he says he can barely distinguish the card much less any letters upon it. We therefore replace the + 6 D. with a + 5 D. over the right eye, and now he tells us he can make out the form of the large letter at the top but he cannot name it. We ask him to look at it quietly for a moment or two while we make the entries of his case in our record book. When we return he tells us the letter is an E although he says it is not sharp and distinct. We have thus secured a slight relaxation of the accommodation and placed the eye in a favorable position for further examination.

This eye is to all intents and purposes, for the time being at least, myopic, and keeping this fact in mind we proceed with the test lenses as we would in any other case of myopia. Where a case is already myopic the fogging system is not applicable, but it is only of value in estimating the refraction of an eye that has some form of hypermetropia, either simple hypermetropia or hypermetropic astigmatism.

We place a $-.50$ D. in front of the convex lenses; this improves vision, makes the E distinct and brings out the letters on the next line. We increase to -1 D. and then to -1.50 D., each change affording a still further improvement in vision, until now the patient can with difficulty discern some of the letters on the No. 30 line. Now we must increase more gradually, and we try a -1.75 D. next. This makes the No. 30 line clear and brings out some of the letters on the No. 20 line. We now advance to -2 D., with which vision is $\frac{20}{20}$ clearly, and a few letters on the No. 15 line can be guessed at.

Now what have we in the trial frame? A $+5$ D. and a -2 D., which by algebraic addition equals $+3$ D., and this represents the amount of hypermetropia present. We examine the left eye in the same way, and find an equal amount of defect.

We have then in this case uncovered 3 D. of the hypermetropia, which the ciliary muscle is constantly overcoming. For all practical purposes this represents the total hypermetropia, although it is barely possible that a strong cycloplegic might show a slight increase. But the measurement of the amount of hypermetropia and the proper glass to prescribe are two different matters. I do not think it wise in this case to attempt to correct the total amount of error.

LET JUDGMENT AND CIRCUMSTANCES DECIDE

There is no hard and fast rule as to just how much should be deducted, and authorities do not agree on this point. We must be guided by the circumstances of each case, the age, occupation, whether glasses have previously been worn, and the condition of the muscle balance. When esophoria is present, we would be justified in prescribing nearly the full correction ; while in the presence of exophoria, we would make a liberal reduction from the total error, for the following reasons : in esophoria the stronger the convex lens, the less the esophoria, and we more nearly approach the normal relation that should exist between the accommodation and convergence. While in exophoria, the stronger convex lens increases the divergence of the visual axes, and hence we give preference to the weaker lens in order to lessen the disparity between accommodation and convergence.

Therefore, before deciding on the glasses to be prescribed, we will inquire into the muscular equilibrium. We will make use of the Maddox rod, which is the most satisfactory all-around muscle test. The rod is placed over the left eye and the patient's attention directed to the light across the room. In answer to our inquiry as to the position of the red streak, the gentleman tells us it is to the left. This is a condition of homonymous diplopia, which we know must be due to an inward turning of the eyes, in other words, to esophoria ; the amount of which is measured by the strength of prism base out that is required to bring the streak back to the light. After trial of a few prisms we find that 8° is the measure of the esophoria.

We are now in position to determine the lenses to be prescribed. Even in view of the decided amount of esophoria, we hardly feel justified in attempting to correct the amount of error, and especially as the young man has never worn glasses before ; but on the other hand the esophoria that is present indicates that the glasses should not be too weak, and therefore taking all the circumstances of the case into consideration, we will order $+ 2.50$ D., which we will advise to be worn constantly. These glasses will probably seem somewhat annoying at first, and we will prepare our patient by telling him that they may cause the ground to appear to slant, and that he may see somewhat better without them, but that he should persevere in their use, and in a week or two they will be entirely satisfactory.

Astigmatism With the Rule

[CLINIC No. 21]

Miss Carrie B., aged twenty-one, complains of headaches.

We notice she is wearing glasses at present, and in answer to our inquiry she tells us she has had them for the past four years. We ask her to let us see the glasses, and we find they are both — .75 D. cyl. axis 180°

The first step in the examination of this case (as of every case of refractive error) is to ascertain the acuteness of vision, which we find to be $\frac{20}{60}$ in each eye. We next find the range of accommodation with each eye without glasses to be from 5" to 11", reading the .50 D. type.

Following the routine method of examination we have advised you to pursue, we now turn to the ophthalmometer to determine the presence or absence of astigmatism.

Examining the right eye, first we focus the instrument so that we obtain a clear image of the mires, and then we revolve the tube to find in which position the long meridian lines are straight and unbroken, which is exactly in the horizontal meridian, and here we find the curvature of the cornea has a value of 45 D. Turning the tube a quarter of a circle, we obtain a reading of 47.50 D. in the vertical meridian. We examine the other eye with exactly the same result.

This, then, is a case of astigmatism "with the rule," the vertical meridian exceeding the horizontal meridian by 2.50 D.

EXPLANATION OF ASTIGMATISM

I have found so much confusion in the minds of my younger students, and I may say also even in the minds of optometrists of considerable experience, about astigmatism with the rule and astigmatism against the rule, that I will give you a short and explicit description of the former as exemplified in the case before us.

You will remember in our studies of the anatomy of the eye, we found that the curvature of the anterior surface of the cornea did not accurately correspond to a section of a sphere, but that the vertical meridian was slightly sharper than the horizontal; this is the condition that is found in the large majority of all eyes.

Now, then, when we have a case of astigmatism in which the vertical meridian, or a meridian on either side of the vertical that is within 45° of it, is more sharply curved than the meridian at right angles to it, we classify the case as one of "astigmatism with the rule."

This simply means that the vertical meridian, or a meridian near to it, presents the sharpest curve and possesses the greater refractive power, and thus corresponds to the usual normal condition of the average pair of eyes. For this reason it is said to be according to the rule, or with the rule. French writers sometimes call this "direct" astigmatism, in contrast to astigmatism against the rule, which they designate as "indirect" astigmatism. Now, gentlemen, this is the whole meaning of astigmatism with the rule, and this is all there is to it.

Sometimes beginners in optometry ask me if hypermetropic astigmatism is not always with the rule, and if myopic astigmatism is not always against the rule. This question occurs to them because they know that hypermetropic astigmatism is usually corrected by a convex cylinder with axis at 90°, and myopic astigmatism by a concave cylinder with axis at 180°. The fact of the matter is that the hypermetropic and myopic astigmatism above mentioned are both with the rule, and it makes no difference whether the astigmatism be hypermetropic or myopic, whether it be simple, compound or mixed, so long as the vertical meridian, or one adjacent to it, has a sharper curvature than the one at right angles to it.

In order to make this perfectly clear, so that no member of the class can go astray on this point, I will make a diagram on the blackboard to illustrare the shape of the cornea in astigmatism with the rule ·

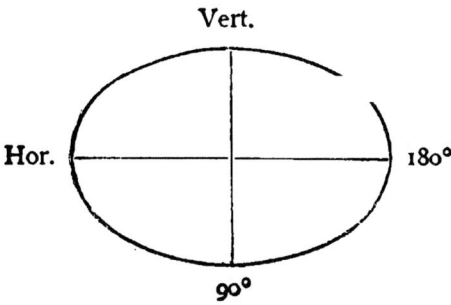

You will see by this diagram that the vertical or 90th meridian is shorter than the horizontal or 180th meridian. Or, in other words, the vertical meridian has a sharper curvature, while the horizontal meridian is flatter. Therefore, the vertical meridian has a greater refractive power than the horizontal, and the focus of the vertical meridian is always in front of the focus of the horizontal meridian.

SIMPLE HYPERMETROPIC ASTIGMATISM

In *simple hypermetropic astigmatism* with the rule, the vertical meridian is emmetropic and the horizontal meridian hypermetropic. Let me call your attention to the fact that the refraction of the vertical meridian is not absolutely increased, but only relatively so, as compared with the horizontal, which is flatter than normal; but still, this corresponds to our definition of astigmatism with the rule, which was that the vertical meridian exceeds the horizontal in curvature. We correct this form of astigmatism by a convex cylinder with axis at 90°, the rationale of which is as follows: the axis of the cylinder being plane, leaves the vertical or emmetropic meridian unaffected; while the horizontal meridian of the cylinder in which its positive refracting power lies, supplements or increases the flat or hypermetropic meridian of the eye and builds it up to normal, so as to equalize the refraction of the two meridians and thus correct the astigmatism.

In *simple myopic astigmatism* with the rule, the emmetropic meridian is now horizontal, while the vertical meridian is myopic. In this case, the refraction of the vertical meridian is not only relatively but absolutely greater than the horizontal, and you have no difficulty in understanding that this condition corresponds to our definition of astigmatism with the rule.

We correct this form of astigmatism by a concave cylinder with axis at 180°, the principle of which is as follows: the axis of the cylinder being plane leaves the horizontal or emmetropic meridian unaffected, while the veritcal meridian of the cylinder in which its diminishing refractive power lies, neutralizes or reduces the excessive refractive power of the vertical or myopic meridian, and makes it the same as the horizontal meridian, thus equalizing the two meridians and thus correcting the astigmatism.

In *compound hypermetropic astigmatism* with the rule, both meridians are flatter than normal and both focus back of the retina, but the horizontal is the flatter of the two and its focus is farther away from the retina than is that of the vertical meridian. The refraction of the vertical meridian is therefore greater than the horizontal, thus corresponding to our definition of astigmatism with the rule.

In *compound myopic astigmatism* with the rule, both meridians have a sharper curvature than normal and both focus in front of the retina, but the vertical meridian has the more pronounced curve and its focus is farther in front of the retina than is that of the horizontal meridian. The refraction of the vertical meridian is obviously greater than the horizontal, and thus without the need of any reasoning is classed as astigmatism with the rule.

MIXED ASTIGMATISM WITH THE RULE

In *mixed astigmatism* with the rule, the conditions are a little more complicated and perhaps not quite so quickly grasped. The vertical meridian is more curved than normal (myopic), and its focus is in front of the retina. The horizontal meridian is flatter than normal (hypermetropic), and its focus is behind the retina. As neither meridian focuses on the retina it cannot be simple astigmatism; but as one is in front and the other back of the retina, it is mixed astigmatism; and as the vertical meridian dominates the horizontal in refractive power, it corresponds to our definition of astigmatism with the rule.

The correction of this form of astigmatism calls for both a convex and a concave cylinder. The convex cylinder is placed with axis at 90°, thus building up the refraction of the horizontal meridian to the normal standard, and leaving the vertical meridian undisturbed. The concave cylinder is placed with axis at 180°, thus reducing the excessive refraction of the vertical meridian to the normal standard, and leaving the horizontal meridian undisturbed. By thus increasing the horizontal and diminishing the vertical, the two meridians are equalized, and the astigmatism thus corrected.

I have shown you by referring to the five forms of regular astigmatism, that it may be with the rule in any or all of its forms; the one thing necessary to make it such is that the curvature of the vertical meridian, or adjacent to it, shall be sharper than the horizontal or the meridian at right angles to it.

I trust I have succeeded in making clear to you the meaning of astigmatism with the rule; and we will now proceed with the examination of our patient. You will remember that we found a corneal astigmatism with the rule of 2.50 D. In one of our former clinics I told you in this form of astigmatism we must deduct .50 D. from the reading of the instrument, and explained to you the reason why; we follow this rule, and now we have 2 D. of astigmatism to be accounted for.

We commence with a .50 D. cylinder, and for reasons with which you are all now familiar, we select a convex and place its axis at 90°. This is accepted, as are also successively a + .75 D. cyl. and a + 1 D. cyl., axis 90°, each in turn affording a still greater improvement in vision. The next stronger lens, + 1.25 D. cyl. axis 90°, was rejected as not being as good as the former lens.

With this + 1 cyl., vision has been raised to $\frac{20}{40}$ clearly. We have reached the limit of acceptance of convex cylinders, but we have not yet raised the vision to normal, nor have we accounted for the 2 D. of astigmatism indicated by the ophthalmometer. This causes us to suspect mixed astigmatism, and we try a — .25 D. cyl. with axis at right angles to the convex cylinder.

This improves vision: we rotate it to 90°, where it is rejected, and then back again to 180°, where it is very much better. This proves that our suspicion of mixed astigmatism is correct, and we cautiously increase the concave cylinder .25 D. at a time, each increase affording a still greater improvement in vision until we reach — 1 D. cyl. axis 180°, with which vision equals $\frac{20}{20}$ clearly, and the ophthalmometric estimate of the astigmatism is all ccounted for.

The left eye is examined in the same way with exactly the came results.

In this case the horizontal meridian shows a deficiency of refractive power, which is corrected by the convex cylinder with axis at 90°, while the vertical meridian shows an excess of refractive power which is corrected by the concave cylinder with axis at 180° The refraction of the horizontal meridian is increased, and that of the vertical meridian diminished, and thus the two meridians are equalized, and the astigmatism neutralized.

COMMON ERROR OF REFRACTIONISTS

It might be profitable for you to consider for a moment the error that was made by the refractionist who fitted this lady's first glasses, which, as I told you, are — .75 D. cyl. axis 180°. I have seen many cases of astigmatism where the cylinders prescribed were concave when they should have been convex. This error is so common, so apt to cause eyestrain instead of relief, and is so inexcusable on the part of an educated optometrist, that I made it the subject of one of our clinics (No. 3), in order to impress it upon our minds, and so that no member of this class would have excuse for such a lapsus.

In this case the myopic meridian is cared for, while the hypermetropic meridian is disregarded. The patient would have received more relief if these conditions had been reversed, because it is the effort on the part of the patient to overcome the hypermetropic meridian (not the myopic) that causes the strain.

This patient's distant vision is poor and she holds her book close, it is true; but in spite of this you must not jump at the conclusion that the defect is entirely myopic. Follow the rule I have so often repeated, to begin the test with convex lenses, either spheres or cylinders, allowing the ciliary muscle a moment's time to accept the + lenses, and do not hastily turn to concaves, not at least until you have used reasonable efforts to have the convexes accepted.

Astigmatism Against the Rule

[CLINIC No. 22]

J. R. S., aged thirty-three years, complains of occipital pain and says his eyes get heavy unless he wears his glasses. Tells us he has been wearing glasses for the past eight years, off and on, which, on examination, we find to be — .75 cyl. ax. 90°.

I want you always to be suspicious of weak concave lenses. I want you almost to forget that there are such lenses in your trial case. I want you to consider the refraction of every case that applies to you as hypermetropic, until positively proven otherwise. I want you to be afraid of concave lenses; they should be marked with the skull and crossbones poison label (just as morphia is labeled in a drug store), so that both alike may be used with caution. They cannot be abolished entirely, but the (would be) label indicates that they must be prescribed with the greatest caution.

When I meet with a case of this kind wearing concave lenses, I instinctively jump to the conclusion that some one has made an error, and I pat myself on the back as being smarter than the other fellow. Of course, I am wrong in my hasty jump sometimes, but then no harm has been done, and I am clearing the tracks so that my examination may be commenced and continued along the proper lines.

A JUSTIFIABLE SUSPICION

This is my feeling as regards the case before us. I am looking on these concave cylinders as being wrong, and I am thinking to myself that they should never have been given to this gentleman, but that he ought to be wearing convexes instead. Now, we will proceed with our examination and see if our conclusions are wrong, or if it's the other fellow that has made the error.

We find the acuteness of vision to be: O. D. $\frac{20}{40}$ clearly; O. S. $\frac{20}{40}$ partly. We hand him the reading card, and note his range of accommodation for the smallest type to be from $4\frac{1}{2}''$ to $20''$.

As I have frequently remarked to you before in a case where the acuteness of vision is reduced to $\frac{20}{30}$ or $\frac{20}{40}$, we always suspect astigmatism, and in order to determine its presence or absence, and in accordance with our routine method of examination, we turn without any further delay to the ophthalmometer.

We locate the principal meridians at 90° and 180°, and find in the right eye that the curvature of the horizontal meridian exceeds that of the vertical by .50 D., and in the left eye by .75 D.

This is contrary to the usual order of things, which is that the excess of curvature generally lies in the vertical meridian, and therefore we see at a glance that this is a case of astigmatism and that it is classified as "against the rule."

At our last clinic we studied a case of astigmatism with the rule, and I explained to you at some length just what was meant by the term. Now, to-day we will consider this case of astigmatism against the rule, and I will endeavor to be just as explicit in my explanation of this condition.

When there is a difference in the curvature of the several meridians of the cornea, and when the excess of curvature lies in the horizontal meridian, or in a meridian either way within 45° of the horizontal, the case is one of astigmatism, classified as against the rule.

DEFINITION OF ASTIGMATISM AGAINST THE RULE

In other words, astigmatism against the rule simply means that the vertical meridian of the cornea is less curved than the horizontal, in contrast to the condition that usually obtains where it is more curved ; and that this deficiency of refraction in the vertical meridian may occur in any form of astigmatism, hypermetropic or myopic, simple, compound or mixed.

In *simple hypermetropic astigmatism against the rule*, the ypermetropia lies in the vertical meridian, while the horizontal meridian is emmetropic. The vertical meridian is less sharply curved than the horizontal, and hence the focus of the vertical meridian is behind the retina, while the focus of the horizontal meridian is on the retina.

While the refraction of the horizontal meridian is normal, yet it is in excess of that of the vertical, and hence corresponds to the definition of astigmatism against the rule which we have given you.

In *compound hypermetropic astigmatism against the rule*, the focus of both meridians is behind the retina, but the vertical meridian being the more hypermetropic, focuses farther back. Now in this case, even though the refraction of both meridians is below normal, yet the horizontal being the least hypermetropic, its refraction exceeds that of the vertical, and therefore causes the case to be classified as against the rule.

In *simple myopic astigmatism against the rule*, the vertical meridian is emmotropic, and focuses on the retina, while the horizontal meridian is myopic and focuses in front of the retina. Therefore the refraction of the horizontal meridian exceeds that of the vertical, and thus classifies the astigmatism as against the rule.

In *compound myopic astigmatism against the rule*, both meridians are myopic and both focus in front of the retina, but the horizontal being the more myopic, focuses farther in front. In this case both meridians possess an excess of refractive power over the normal, but the horizontal, having the greater excess, corresponds to our definition of astigmatism against the rule.

In *mixed astigmatism against the rule*, the vertical meridian is hypermetropic and focuses back of the retina, while the horizontal meridian is myopic and focuses in front of the retina. The excess of curvature therefore lies in the horizontal meridian, and as this is contrary to the usual order of things, the case is classed as one of astigmatism against the rule.

THE SHAPE OF THE CORNEA

In all of these five forms of regular astigmatism, the shape of the cornea as viewed from the front, is that of an oval, with the longest meridian in the vertical direction.

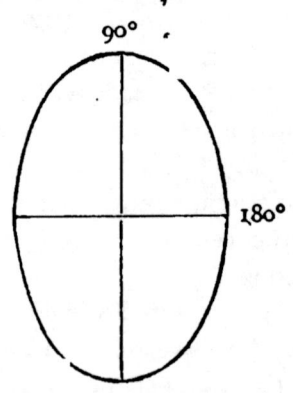

I will make a diagram on the blackboard to represent this condition, and will ask you to call to mind each form of astigmatism and compare it with the diagram. You will understand that the longer the curve, the flatter it will be and the less its refractive power; while the shorter the curve, the sharper it will be and the greater its refraetive power.

In simple hypermetropic astigmatism the vertical meridian is longer and flatter, while the horizontal meridian is emmetropic, and exceeds the vertical because the latter is below par.

In compound hypermetropic astigmatism, both meridians are hypermetropic, the horizontal being less defective than the vertical, and hence the latter has the shorter and sharper curve.

In simple myopic astigmatism, the vertical meridian now is normal, and the horizontal myopic; therefore the latter has the shorter and sharper curve.

In compound myopic astigmatism, both meridians are myopic, but as the vertical is less myopic than the horizontal, the latter has the shorter and sharper curve.

In mixed astigmatism, where the vertical meridian is hypermetropie and the horizontal myopic, the latter has the shorter and sharper curve.

This demonstrates that astigmatism may be against the rule in any and all of its five forms. So long as the vertical meridian has the longer and flatter curve, and the horizontal the shorter and sharper curve, the astigmatism is against the rule.

I have gone thus minutely into the point of astigmatism against the rule, because I have found so many students who have confused ideas about it. A great many of our text books make but slight mention of it, and attach but little importance to it, but I have found it of great value in instructing students and making the subject clear to you.

MENTAL PICTURE OF THE EYE

The optometrist is thus enabled to picture in his mind's eye the condition of the curvatures of the cornea in any particular case of astigmatism, and the location of the focal points of each meridian in relation to the retina, and this certainly is of advantage to him in the adjustment of glasses for the correction of the particular defect which he finds. It makes him think of the eye under observation, rather than of some obstruse rule from a text book.

How can you quickly and definitely diagnose a case of astigmatism against the rule? By the opthalmometer, showing that the mires overlap in the horizontal meridian, or at least are closer together than in the vertical meridian. If you will refer to the diagram on the blackboard you will see the reason for this approximation of the images in the horizontal meridian in the shortened radius of curvature in this direction.

Consequently, when you are looking at an eye through the ophthalmometer, if on turning from the primary to the secondary position, you see the mires separate, you know at once you have a case of astigmatism against the rule, because this separation indicates a longer and flatter curve vertically, which necessarily means a shorter and sharper curve horizontally in contrast.

THE TEST CASE EXAMINATION

We are now ready to proceed with our test case examination. As the ophthalmometer has disclosed the presence of astigmatism, we commence at once with cylinders, and as the instrument has shown the defect to be against the rule, we place the axis of the cylinder in the horizonal position.

Our patient tells us that + .50 cyl. axis 180° produces a noticeable improvement in vision. We hold a + .25 cylinder over the + .50 cylinder, expecting it to be accepted, because, as I have already explained to you at a previous clinic, in astigmatism against the rule, we add .50 D. to the reading of the ophthalmometer, which in this case in this eye would show 1 D. of astigmatism. But to my surprise this additional + .25 D. is rejected.

In such a case you must try a concave cylinder with axis placed at right angles. I therefore place a — .25 D. cylinder with axis at 90°, and at once our patient's face lights up and he says the letters are very much plainer, and he can now name some of the letters on the No. 20 line. We replace this — .25 cylinder with a — 50 D. cylinder axis in same position, which renders the No. 20 line perfectly clear and legible.

This proves the case to be one of mixed astigmatism, the vertical meridian being .50 D. less than normal and the horizontal meridian .50 D. above normal, the difference between the two meridians being 1 D., as indicated by the ophthalmometer.

The vision of the left eye is not as good as the right, and the ophthalmometer discloses a greater degree of astigmatism. We place + .50 D. cylinder with axis at 180°, which is at once accepted; we hold a + .25 D. cylinder in front of it, which is still better, and then a + .50 D. cylinder, which affords still greater improvement. This is equivalent to a + 1 D. cylinder, which we place in the trial frame with axis at 180°, with which vision equals $\frac{20}{20}$ almost.

We now place a + .25 D. cylinder in front of it, and this is promptly rejected. But as the ophthalmometer indicates a greater degree of astigmatism than 1 D., we try a — .25 D. cylinder with axis at right angles. This is immediately accepted as a great improvement and makes vision $\frac{20}{20}$, every letter being clear and legible. This eye also presents a case of mixed astigmatism, the difference between the two meridians being 1.25, which corresponds exactly to the indications of the ophthalmometer.

TO ASSURE CORRECT DIAGNOSIS

In the early part of our lecture I called your attention to the concave cylinders the patient has been wearing, and told you they were probably wrong, and it was likely they should be convex instead. Now, since we have completed our examination you will see that I was not altogether in the right in my surmise, nor was the other fellow altogether in the wrong in his prescription.

It is perhaps not so much of an error to prescribe concave cylinders in a case of mixed astigmatism as in a case of hypermetropic astigmatism. Perhaps you might be inclined to regard it as only half an error, because one meridian is corrected and the other not. But at the same time it is an error, and especially so in the left eye, where the hypermetropic meridian is so much greater than the myopic. If only one meridian was to be corrected, it would have been much better to correct the hypermetropic meridian instead of the myopic, as this would have lessened the tax on the accommodation.

Now this error could have been avoided if the first optometrist had followed out the method of examination which I have taught you in the use of the test case, and that is always to commence with convex lenses, either spheres or cylinders. In this case the convex cylinders were accepted without any hesitation, and there was no need to turn to concave cylinders except in combination with the convexes.

In completing our examination, we use the ophthalmoscope, which reveals nothing abnormal in the fundus except a slight congestion, probably due to strain.

We also test the muscular balance by means of the Maddox rod, and find 1° of esophoria and 1° of left hypherphoria. We will not attempt to correct these muscular anomalies as yet, until we see what relief will be obtained from the correction of the mixed astigmatism. Therefore, we will order (as a result of the transposition of the cross-cylinders)

O. D. + .50 D. sph. ◯ — 1 D. cyl. axis 90°
O. S. + 1 D. sph. ◯ — 1.25 D. cyl. axis 90°

Lenticular Astigmatism

[CLINIC NO. 23]

Mrs. H. H., aged thirty years, in good health apparently, complains of headaches and pains in the eyes, especially after using them for close work.

We find the acuteness of vision to be $\frac{20}{50}$, both eyes being alike. Range of accommodation for Jaeger No. 1 from 7 to 15 inches.

In accordance with our custom, we ask the patient to be seated at the ophthalmometer. We find the primary position and meridian of least curvature to be exactly at 180°, and at 90° an excess of refracting power amounting to 2.50 D. This classes the case as one of astigmatism with the rule.

We now turn to the trial case examination and commence with + .50 D. cyl. axis at 90°. This is rejected by the patient as being decidedly worse. In this case, if the ophthalmometric readings are to be depended upon, a minus cylinder with an axis horizontal is indicated. We try a — .50 D. cyl. axis 180°, which is accepted by the patient as being noticeably better.

We increase — .25 D. at a time, each change affording a still greater improvement in vision, until we reach — 1.25 D. cyl. axis 180°, with which vision equals $\frac{20}{20}$ clearly. We must not order a stronger concave lens than absolutely necessary, and as these — 1.25 cylinders raise vision to normal, we have no justification in trying stronger ones.

AN INTERESTING DISCREPANCY

Now, when we come to consider this case, we find a discrepancy between the amount of astigmatism indicated by the objective test (ophthalmometer) and that indicated by the subjective test (trial lenses).

You will perhaps remember at one of our previous clinics, when we had a case of astigmatism with the rule before us, I told you we must deduct .50 D. from the reading of the ophthalmometer, and explained to you the reason therefor. By making such deduction in this case, we can find that the ophthalmometer indicates 2 D. of astigmatism, while the test lenses disclose only 1.25 D. of astigmatism, a difference of .75 D.

This case, therefore, must be considered as somewhat exceptional. The early workers with the ophthalmometer believed, in the majority of cases, that the total astigmatism and the corneal astigmatism were the same. But a further and more extended use of the instrument soon established the fact that there is a difference between the total astigmatism and the corneal astigmatism of about .50 D. or .75 D.

When the astigmatism is with the rule, the total error is found to be .50 D. less than the corneal, while in astigmatism against the rule, the total defect is found to be .50 D. more than the corneal. The axis of the total astigmatism usually coincides with that of the corneal.

This difference between the total and corneal astigmatism led to the enunciation of the rule, with which you are all now familiar, to add or subtract .50 D., as the case was against or with the rule, and thus make the objective and subjective examinations to correspond. In fact, when the reading of the ophthalmometer is verified by the test case examination, you may feel safe in prescribing the glasses thus indicated.

But sometimes there is a discrepancy between the results of the two methods, as in the case before us, and then it becomes a matter of interest and importance to investigate the possible cause of such difference.

LENTICULAR ASTIGMATISM

Among the several causes advanced by various authorities, the most important are abnormal lenticular astigmatism, and astigmatism of the posterior surface of the cornea. It is to the first of these that I wish to direct your attention to-day

The seat of astigmatism may be either extra-ocular (when on the anterior surface of the cornea), or intra-ocular (when located in any of the refracting media, more especially in the crystalline lens). Therefore, the total astigmatism is made up of the sum of the corneal and lenticular.

It is but a comparatively short time since Javal and other observers established the fact that there is an astigmatism of the crystalline lens, and that it amounts, as a rule, to .50 D. or .75 D. This may be called the normal astigmatism of the lens, just as we have the same amount of astigmatism normally present in the cornea. In the latter case the excess of curvature is in the vertical meridian, while in the former the excess is in the horizontal meridian, and as the departure from normal is about the same in each case, one tends to neutralize the other.

Lenticular Astigmatism

In corneal astigmatism with the rule, there is usually associated a lenticular astigmatism of .50 D. to .75 D. in the *same* meridian, but of an *opposite* kind, thereby neutralizing that amount of corneal astigmatism.

In hypermetropic astigmatism with the rule, it is the horizontal meridian of the cornea that is flatter or hypermetropic, while it is the vertical meridian of the crystalline lens that is similarly affected. In other words, the flatter meridian of the cornea corresponds to the more convex meridian of the lens, and the more convex meridian of the cornea to the flatter meridian of the lens.

EXPLANATORY DIAGRAMS

In the non-astigmatic eye, the vertical meridian of the cornea is + .50 D., while the horizontal meridian of the cornea is emmetropic. The vertical meridian of the crystalline lens is emmetropic, while its horizontal meridian is + .50 D.

I will make a diagram on the blackboard, which may perhaps serve to make my meaning clearer:

	VERTICAL	HORIZONTAL
Cornea . . .	+ .50 D.	∞
Crystalline .	∞	+ .50 D.
	+ .50 D.	+ .50 D.

The refraction of the two meridians is thus equalized, the astigmatism is eliminated, and the case reduced to one of simple hypermetropia.

Or the conditions present may be represented by the following diagram:

	VERTICAL	HORIZONTAL
Cornea . . .	+ .50 D.	∞
Crystalline . .	− .50 D.	∞
	∞	∞

Here the vertical meridian of the cornea is + .50 D. above normal, while the vertical meridian of the crystalline is — .50 D. below normal, resulting in an elimination of the astigmatism and establishment of a condition of emmetropia.

Another diagram will serve to illustrate the conditions in an astigmatic eye, with the rule.

	VERTICAL	HORIZONTAL
Cornea . . .	+ 1.50 D.	∞
Crystalline . .	— .50. D.	∞
	+ 1 D.	∞

In this case the refraction of the vertical meridian is reduced 50 D., but the astigmatism still remains, being lessened only in amount.

In astigmatism against the rule, the conditions present may be illustrated by the following diagram

	VERTICAL	HORIZONTAL
Cornea . . .	∞	+ .50 D.
Crystalline . .	∞	+ .50 D.
	∞	+ 1 D.

In this case, the excess of curvature in the cornea lies in the horizontal meridian, as it does also in the crystalline lens, and therefore as the astigmatism is in the *same* meridian and of the *same* kind, the total amount of defect is increased.

CAUSES OF LENTICULAR ASTIGMATISM

The question may occur to you as to how regular lenticular astigmatism is caused. It may be produced by an oblique position of the lens, by a slight displacement or by unequal curvatures of its surfaces. It is possible, also, to have a dynamic lenticular astigmatism as the result of an unequal contraction of the ciliary muscle.

If it was not so nearly covered by the iris, the position of the crystalline lens could be easily determined. A case has been reported of complete absence of the iris (aniridia), in which the crystalline lens could be plainly seen. When patient first came under observation, the lens was vertical and not displaced at all, but in the course of a year or two it became luxated upward about 1½ mm., and the upper edge tilted slightly backward. On this account the total astigmatism was increased 1.50 D., on account of the increased astigmatism of the lens, while that of the cornea remained unchanged.

It is a fact beyond dispute, that in nearly all cases of astigmatism with the rule, and they make up the great majority of all cases of astigmatism, the amount of corneal astigmatism shown by the objective method is lessened by the subjective examination, undoubtedly by a neutralizing lenticular astigmatism.

Those cases of dynamic astigmatism of the lens, produced by an unequal contraction of the ciliary muscle, are revealed by paralyzing the accommodation by atropine. This leaves the corneal astigmatism unchanged, and at the same time it lessens or removes the lenticular astigmatism.

The result is that now the corneal astigmatism calls for a cylindrical lens, while previously such lens was rejected because of the neutralizing effect of the dynamic lenticular astigmatism, which has been removed by the action of the mydratic.

The conditions we have been considering are where the lenticular astigmatism is present to the normal amount of .50 D. or .75 D., diminishing or increasing the corneal astigmatism by that amount, as it is with or against the rule. I have already explained to you how the readings of the ophthalmometer are to be interpreted and modified in accordance with these well-established facts.

AN EXCEPTIONAL CASE

But in the case before us I have made these deductions, and in spite of this the results of the objective and subjective tests do not agree. The ophthalmometer showed that the refraction of the horizontal meridian was 2.50 D. less than the vertical. I deduct the customary .50 D. of lenticular astigmatism, leaving 2 D. of astigmatism with the rule, but the subjective examination by the test case showed only 1.25 D. of astigmatism, a difference of .75 D.

This case, therefore, is unusual or exceptional. What inference can we draw from these facts, or what explanation can be offered for this difference? Most likely that the lenticular astigmatism, instead of being .50 D. or 75 D., as usual, amounts to 1.25 D., the excess of curvature being in the horizontal meridian.

I will make a diagram on the blackboard to demonstrate the conditions present in this case :

	VERTICAL	HORIZONTAL
Cornea . . .	+ 2.50 D.	∞
Cyrstalline . .	1.25 D.	∞
	+ 1.25 D.	∞

If you will consider this diagram, you will see that the astigmatism is in the same meridian, but of opposite kind. The cornea shows an excess of 2.50 D. in the vertical meridian, which means a myopia of this amount in the vertical meridian. The crystalline shows a deficiency of 1.25 D. in the vertical meridian, which means a hypermetropia of this amount in this meridian. The hypermetropia neutralizes a portion of the myopia, and leaves 1.25 D. of defect in this meridian, to be corrected by a concave cylinder with axis at right angles.

In both the cornea and the crystalline, the horizontal meridian is assumed to be of the normal curvature or emmetropic.

This is one case where the lenticular astigmatism exceeds the usual amount of .50 D. or .75 D., and, of course, there are others. The lenticular astigmatism may amount to 2 D., and in rare exceptions to even more. I have knowledge of one case being reported where it reached 7.50 D., but such a case is extraordinary.

In the practice of optometry you must be prepared for exceptional cases. Your text-books and your teachers can lay down only general rules for your guidance, and general principles by means of which you can gain an orderly knowledge of the science. But when you come to put these principles into practice, you must not be surprised when you meet with cases where there are variations from the conditions which you have been taught, sometimes slight and sometimes very marked.

AVOID A HASTY CONCLUSION

When you find a discrepancy, you must not too hastily conclude that it is due to lenticular astigmatism. For instance, you may have made an error in observation, due to a poor light, restlessness of the patient or an improper position of the head in the headrest. Or an accumulation of tears will cause an incorrect estimate to be made. If the tears stand in the groove between the lower lid and the ball, the ophthalmometer may read astigmatism against the rule, where there is actually astigmatism with the rule. You can easily understand how tears may alter the refractive power of the eye, because they form the first refractive surface which the light strikes. There is always a thin layer of tears covering the cornea, but ordinarily it is so slight and in such close contact, that they have no perceptible influence. But when they collect in excess and encroach upon the lower half of the cornea, they may very readily alter the readings of the ophthalmometer.

Sometimes the instrument may be at fault, as, for instance, an imperfect adjustment of the bi-refractive prism in the telescope : on account of which, no matter how carefully the primary position is obtained, the secondary position would show false results.

Or if the arc was not adjusted to exactly coincide with the line of doubling of the prism, it would be impossible to make the images line in any position of the instrument. Or the prisms and lenses of the ophthalmometer may be made of imperfect material.

But these errors can be overcome, and no matter how perfect the result of the objective examination, it must be verified or disproved by a subjective examination, which if properly made, is after all the court of last resort. Both methods of examination are important and neither can be dispensed with, but their results must always be interpreted with common sense, and free from adherence to any fixed rules or notions.

A Case of High Myopia

[CLINIC NO. 24]

E. C. is a school girl, aged eleven years. She holds everything very close to her eyes and is unable to see the blackboard in school. Her father tells us she has been wearing glasses for the past six years. We neutralize them and find them to be — 5 D. spheres. This girl, as you see, is a delicate-looking child, but makes no complaint other than her inability to see clearly.

As might be expected, she is unable, without glasses, to see any of the letters on the test card hanging across the room. We ask her to approach the card slowly, and when about eight feet from it she is able to name the large letter at the top. We therefore record her visual acuity as $\frac{8}{200}$.

We ask her to return to her chair, and testing each eye separately with concave lenses, we find for right eye — 12 D. affords a vision of $\frac{20}{100}$, and for the left eye — 18 D. the same acuteness of vision.

This, then, is a case of high myopia, a very serious condition of refraction for the patient, and possessing many points of interest to the optometrist.

This is the one error of refraction in which an ophthalmoscopic examination is of the highest importance, in order to determine the condition of the interior of the eye, and the presence of the myopic crescent, choroidal atrophy, macular disease, opacity of the vitreous humor or other accompaniments of staphyloma.

Myopia, as you all know, is ordinarily due to excessive length of the antero-posterior diameter of the eye, and we are able to estimate pretty accurately the amount of lengthening by the number of the lens required to correct the myopia, according to the following standard, that every 3 D. lens represents very closely one millimeter of lengthening of the antero-posterior diameter of the eye.

In this little girl's case, the lenses just accepted would indicate 4 mm. of lengthening in the right eye, and 6 mm. of lengthening in the left eye.

Inasmuch as hypermetropia is regarded as an imperfectly developed condition, so it might seem fair to look upon the myopic eye as one that has undergone excessive development; but the fact of the matter is that in the vast majority of cases, the excessive length is due, not so much to overgrowth as to stretching and distention of the ocular coats.

Much discussion has occurred as to how this stretching has been caused, and the various theories that have been advanced are divisible into two general classes:

1. Those which attribute the deleterious effect to the prolonged exercise of the accommodation.

2. Those which attribute this effect to the convergence.

The supporters of the first theory argue that the intra-ocular pressure is increased during accommodation, and that distention of the sclerotic is due to the long continuance of this abnormal pressure; that the act of accommodation causes traction to be exerted upon the choroid, thereby giving rise to chronic inflammatory changes with subsequent atrophy and thinning of the choroid and sclerotic; that spasm of accommodation is an important factor in the causation of myopia.

All of these have been denied, and a potent argument against the accommodation theory lies in the fact that there is no increase of refraction in those eyes where the accommodative effort is the greatest, namely in hypermetropia.

INFLUENCE OF CONVERGENCE

I am inclined to regard the influence of the convergence upon the shape of the eyeball as of more importance than the accommodation. When the internal recti are strongly contracted in the function of convergence, the external recti closely bind the outer halves of the balls, and at the same time the two obliques must increase their effort in order to prevent the balls from sinking back into the orbits. The pressure upon the eyes is thus increased and traction made upon the posterior part of the sclerotic by the oblique muscles.

These efforts of accommodation and convergence are common to all who use their eyes for close work, but as a matter of fact only a certain proportion of them become myopic, and therefore it is necessary to assume the assistance of a predisposing cause in those eyes which do become elongated.

A large broad skull and a great interpupillary distance, render convergence more difficult and thus present a predisposing element.

Heredity is another predisposing factor to myopia ; this does not mean that the babe is born myopic. On the contrary, the eye is very likely to be hypermetropic at birth, but what the child inherits is a tendency to myopia on account of weakness of the coats of the eye or a subnormal resisting power of the sclerotic. Such eyes give way under a strain that would be harmless to a strong-coated eye.

So great is the elongation in the higher grades of myopia that the sclerotic is reduced to extreme thinness, and on account of the underlying choroid it assumes a bluish tint. This protrusion backward of the myopic eye is termed posterior staphyloma.

The conus is a whitish crescent (known as the myopic crescent) found at the border of the optic nerve ; or instead of being crescent shaped, it may entirely surround the optic disk. This is due to stretching and atrophy of the choroid, allowing the white sclerotic to show through. The presence of a conus may be regarded as evidence of congenital deficiency in the resisting power of the sclerotic.

TESTING THE CASE

We will now return to our case and make a more careful and thorough test with the trial lenses ; but before doing so we will call the ophthalmometer to our aid to determine the presence or absence of astigmatism. This instrument shows an excess of curvature in the vertical meridian of 2.50 D. in the right eye, and 1.50 D. in the left. This indicates astigmatism with the rule, and if a concave cylinder is called for the axis would be placed at 180°.

We place a — 10 D. in front of the right eye, and hold before it alternately a — 1 D. sphere and a — 1 D. cylinder with axis at 180°. The sphere is preferred as affording the better vision. We now place a — 11 D. in the trial frame and repeat the process. We keep on along these lines until finally we get — 13 D. $\bigcirc$ — 2 cyl. axis 180° as the best combination we can find, with which our little patient is able to name some of the letters on the No. 60 line.

We repeat the test in the same way with the left eye, taking plenty of time and exercising great patience, with the following result : 16 D. sph. $\bigcirc$ — 1 cyl. axis 180°, with which vision equals $\frac{20}{80}$.

This is certainly a high degree of myopia, and we must be on our guard not to tax the eye by giving the glasses too strong. Therefore we will slightly reduce the sphere and order as follows:

O. D. — 12 D. S. ◯ — 2 D. cyl. axis 180°
O. S. — 14 D. S. ◯ — 1 D. cyl. axis 180°.

These glasses can be worn only for distance, while for reading we will have to modify the glasses according to the following rule: Substract from the glasses representing the full measure of the defect those glasses the focus of which represents the distance at which the patient desires to read or work.

Now the usual reading or working distance is 13 inches, and the glass representing this distance is 3 D. and therefore this is the amount that is subtracted from the distance lenses, which would make the prescription for reading glasses as follows:

O. D. — 9 D. S. ◯ — 2 D. cyl. axis 180°.
O. S. — 11 D. S. ◯ — 1 D. cyl. axis 180°.

SUGGESTIONS FOR PARENTS

A child at this age cannot be expected to change her glasses from time to time as she may be looking near or far, and therefore we will suggest to her parents that these reading glasses be worn constantly while at school and at play, and that the distance glasses be reserved for the church or theater, or when she makes some special visit where her best vision may be desirable.

We will make our usual test of the muscle balance by means of the Maddox rod, and find 15° esophoria and 1½° left hyperphoria. This is somewhat unexpected, as ordinarily esophoria is associated with hypermetropia, while a myopic condition of refraction gives rise to a divergence of the visual axes, which may show itself simply as an exophoria or as an actual divergent squint.

This departure from parallelism of the visual axes is due to the disturbance of the relations that should normally exist between accommodation and convergence. In myopia there is little or no need of the accommodation in near vision, but the convergence must be used the same as in an emmetropic eye. The convergence is then used in excess of the accommodation, which leads to fatigue of the internal recti and finally insufficiency.

Or looking at the matter in another light: accommodation and convergence are used in equal proportion in the normal eye, and each function receives the same nervous impulse, causing an equal effort. Now, in myopia the divergent rays from near objects are focused on the retina with little or no accommodative effort. For this reason there is no call for innervation of the ciliary muscle, and consequently the innervation of the internal recti is lessened or checked to an equal degree. Thus relaxation of these muscles is produced, and the eye under the control of the stronger external recti turns outward.

In this case, however, instead there is a decided convergence of the visual axes, showing that the normal relation that should exist between the accommodation and convergence has been destroyed or disturbed. We feel as if some attention should be given to this imbalance of the muscles, but on account of the strong concave curves that are called for, we hesitate to order prisms in combination. How, then, can we obtain the desired prismatic action? By decentration.

DECENTRATION

As soon as we look through a lens at any place except its actual optical center, the prismatic effect of the lens is brought into action. If the lens is weak and the curvatures slight, the prismatic effect is inappreciable; but with the increase in power and curvature of the lens, there is a corresponding increase in prismatic action. In the case before us, where the lenses required are so strong, we can get very considerable prismatic power by decentration.

Now, is there any rule to guide us in this matter? Of course there must be, in order that we may know exactly what we are doing.

In the first place, a concave lens may be considered as made up of indefinite number of prisms with their bases out. If such a lens is decentered outward, we get the effect of a prism base in; if decentered inward, a prism base out. In this case where esophoria is present, we want a prism base out, and therefore we must order the lenses decentered inward. Now the question occurs, how much shall they be decentered?

The rule is that for every decentration of one centimeter there will be as many degrees of prism as there are diopters in the lens. This means that a 1 D. lens decentered 1 cm. would produce a prismatic effect of 1°, and a 3 D. decentered 1 cm. a prismatic effect of 3°.

But on account of the small size of uncut lenses, a decentration of 3 mm is all that is possible, and therefore we had better word our rule in accordance therewith, as follows : a 1 D. lens decentered 1 mm. produces a prismatic effect of .1 D ($\frac{1}{10}$), and a 3 D lens decentered 1 mm. a prismatic effect of .3 D. ($\frac{3}{10}$). Inasmuch as the size of the lens limits the decentration to 3 mm., the amount of prismatic power it is possible for us to produce is .3° ($\frac{3}{10}$) for every 1 D. of refractive power.

In this case, where we have ordered — 12 D. for the right eye and — 14 D. for the left, we can produce 3.6° and 4.2° of prismatic power, respectively, or a total of 7.8°, almost 8°. As this is about one-half the amount of esophoria, and as this is about as much as we usually attempt to correct, you can easily see that a decentration in this case is of great practical value, and we will therefore order the lenses decentered inwards 3 mm.

Now, there was also 1½° of left hyperphoria, which we can correct in the same way. For this purpose we want the effect of a 1° prism base down over the left eye. We get the effect of a prism base down by ordering the lens decentered upwards, and if it is decentered 1 mm. we get a prismatic effect of 1.4°, which is just about what we want.

REMOVAL OF CRYSTALLINE LENS

In considering the treatment of a case of high myopia like the one before us, the question of the removal of the crystalline lens, which has recently been advocated, presents itself. Theoretically, this is a beautiful plan of treatment, as in this way the excessive refraction can be very materially reduced, but it does not appeal very strongly to the average patient, and therefore it is scarcely likely to become very popular.

Hundreds of cases of removal of the crystalline lens have been reported in Europe and especially in Germany. So far, this country has furnished very few cases ; one reason for which may be found in the fact that we give more attention to the careful and painstaking correction of errors of refraction than do the European optometrists.

While on this subject, it is interesting to consider the amount of myopia that would be suitable for operation, The least degree of myopia in which extraction of the lens is permissible is about 12 D. Of course, the higher the degree, the greater might be considered the need for operation. But at the same time, in the higher degrees of myopia, the accompanying posterior staphyloma impairs the integrity of the coats of the eye so greatly that the operation is attended with serious risks, such as hemorrhage and detachment of the retina.

A condition approximating emmetropia may result from extraction of the lens in myopia, varying from 12 D. to 20 D. This case would be included in this class, if the conditions demanded operation ; but with a vision as good as this child enjoys, we would not think of advising operative procedure.

As you have noticed the best vision we have been able to obtain in this case is $\frac{20}{60}$ partly, and this leads me to say that in the higher grades of myopia, it is impossible to raise vision to normal by any lens. This, perhaps, may be comforting knowledge to some of you who have vainly tried to find a lens or a combination of lenses that would afford a vision of $\frac{20}{20}$ in the highly myopic cases that have come under your care.

There are two reasons for this ·
1. The impaired integrity of the retina.
2. The diminishing effect of strong concave lenses.

Either one of these would suffice to account for the lessened vision, while the two together only serve to make it more pronounced.

You can easily understand that the great bulging and stretching of the coats of the eye, causes the rods and cones of the retina to be separated. The diminution in the size of the image by a strong concave lens is very marked, therefore this smaller image impresses fewer of the rods and cones, and there can be no wonder that the vision is not capable of being raised to normal. Indeed, the wonder rather is that we are able to afford as good a vision as we do with the strong concave lenses we are compelled to prescribe.

Keratoconus or Conical Cornea

[CLINIC No. 25]

J. E. B., forty-six years of age, complains of headache and indistinctness of vision. He says he has been wearing glasses for the last eleven years, and that previously his vision had been good. A glance at his glasses shows them to be strong cylinders.

On seating him at the usual distance from the test-card and removing his glasses, we find he is unable to name the large letter at the top of the card. In a case of impaired vision like this, it is desirable to determine whether the refraction of the eye is at fault or whether the defective vision is dependent upon disease. In the pin-hole disk, we fortunately possess a method by which this point can be easily and quickly determined.

We will therefore make use of this test in this case. You will notice that I do not place the pin-hole disk in the trial frame, because in so many cases I have found that the patient has difficulty in finding the opening and his answer is that he can see nothing.

Instead, I prefer to give the disk to the patient to hold in his hand, while he covers with his other hand the other eye. In this way he has the disk entirely under his control, and he can easily move it around until he gets it directly in the visual line. In this way there is admitted into the eye a small pencil of light, which, passing through the axis of the refractive system of the eye, forms a clearly-defined image on the retina, in spite of any errors of refraction that may be present.

Therefore, if the pin-hole improves vision, we know that the refractive system of the eye is at fault, and that a similar or greater improvement in vision can be expected from glasses.

If, on the contrary, there is no improvement in vision by look-through the pin-hole disk, there must be some diseased condition which is not remediable by glasses, so that no matter how perfect the image that is formed in the eye, it cannot be perceived or transmitted to the brain, and thus the case is classed as one beyond the province of the optician.

USE OF THE PIN-HOLE DISK

I would advise you in cases of defective vision to use the pin-hole disk at once, and discover early in the examination whether or not the case is one of refractive error, and you will thus save much valuable time which might be spent in trying to fit a case that could not be helped by glasses.

Any one of you gentlemen can demonstrate for himself the effect of the pin-hole disk. Take from your trial case a strong convex lens, hold it close to your eye and look through it at some distant object, as the test-card. The letters will all be indistinct, in fact will be entirely blotted out; now place the pin-hole disk in front of the lens, when you will find its power is destroyed and vision is restored to normal. In like manner, the effect of a strong concave lens will be neutralized, as will also the effect of cylinders, whether convex or concave.

If the damaging effect on vision caused by imposition of strong lenses can be thus destroyed by the use of the pin-hole disk, you can easily understood how the impaired vision of refractive errors can be improved by the same means.

We now hand the pin-hole disk to this patient, and ask him to use it in the way we have just described. The result, as he tells us, is O. D. $\frac{20}{100}$, O. S. $\frac{20}{80}$. This is not very encouraging, but at the same time it demonstrates the possibility of some improvement in vision by means of carefully-adjusted lenses.

We will now make use of the ophthalmometer. I find great distortion in the shape of the mires, and I am unable to focus them sharply. This shows greatly irregularity in the curvatures of the cornea, and from my previous experience in similar cases, I recognize here a case of keratoconus or conical cornea.

I am unable to get the primary position of the mires, or the meridians of least and greatest refraction or the amount of overlapping in the latter; the most that I can do with the ophthalmometer is to locate one of the principal meridians in each eye, that for the right eye being at 75° and for the left eye at 105°.

OPHTHALMOMETER OF GREAL VALUE

Before the days of the ophthalmometer, cases like this with irregular curvature of the cornea, presented the greatest difficulties in the fitting of glasses, but the perfected instrument of the present day does much towards overcoming these difficulties.

The first glance through the ophthalmometer reveals the irregularity of surface, and at the same time we gain information about any regular astigmatism that may be present. In this case the ophthalmometer shows the presence of regular astigmatism in addition to the irregular. The distortion of the images of the mires is such a delicate test, that the slightest irregularity of the surfaces of the cornea is detected. The ophthalmometer is also of the greatest value in finding the most regular part of the cornea.

In many of the cases of conical cornea, the astigmatism is of such a high degree that the ophthalmometer, as now constructed, is only capable of giving the relative difference of the dioptric power of the two chief meridians of the cornea, and not absolute and exact measurements. Nevertheless, in all ordinary cases and for all practical purposes, the instrument is accurate enough. In the exceptional cases, the difference in the curvature of the two chief meridians can be approximated, as can also the position of the two meridians.

We will now use the ophthalmoscope in this case, and we find the shadow crescent of conical cornea beautifully shown in each eye. In the right eye the shadow is so pronounced as to suggest an opacity of the lens, but as we examine by oblique illumination the lens and cornea show perfectly clear.

The details of the fundus are but indistinctly seen, with either the direct or indirect method. The optic disk is whitish, and is long and narrow vertically, but I can see only parts of it at a time, the blood vessels and background changing with each movement of the eye or the ophthalmoscope. There are no opacities in the vitreous humor.

The ophthalmoscope is of no value in the estimation of the refraction, while the retinoscope is utterly useless, as we cannot get any definite reflections or movements. The subjective test with the clock dial is altogether unsatisfactory. The value of the ophthalmometer is greatly limited, but it has pointed out to us the location of one of the principal meridians. This narrows the measurement of the error of refraction down to the subjective tests with the test case and trial lenses.

TESTING WITH TRIAL LENSES

We will commence the test with the right eye. The ophthalmometer has indicated the location of one of the principal meridians at 75°, but has given us no evidence as to whether it is the meridian of least or greatest refraction, and hence we will have to start with the test lenses somewhat empirically.

We place a + 1 D. cylinder in the trial frame with axis at 75°. The result is negative, or if anything a still greater dimness of vision. We then rotate the cylinder, and the patient chooses 165° as the best position for the axis. This indicates a case of astigmatism against the rule, the vertical meridian (or 15° from it) being flatter or hypermetropic. We add a .50 D. cylinder with axis in same position, and still another .50 D. cyl., both of which are accepted, and with this + 2 D. cylinder we have reached the limit of convex acceptance, but so far have produced but little improvement in vision.

WHAT THE EXAMINATION INDICATES

Presuming the astigmatism to be of much higher degree, we place a — 1 D. cylinder over the convex cylinder, with axes at right angles. This makes the letters considerably brighter, and the acceptance of this concave cylinder with axis at 90° (or within 15°) shows the horizontal meridian to be convex or myopic, and indicates astigmatism against the rule ; or, taken in connection with the convex cylinder, a case of mixed astigmatism against the rule.

We increase the concave cylinder 1 D. at a time, each change producing a greater improvement in vision until we reach — 6 D. cylinder, which is the strongest cylinder in our trial case, but without affording very satisfactory vision. It is very unusual to meet with astigmatism of a higher degree, and therefore stronger cylinders are but seldom called for.

Fortunately we have a three-cell trial frame, which holds the two cylinders so far selected, and permits of an additional concave cylinder, which of course is placed with its axis in the same position. We add a — 1 D. cylinder, which affords considerable improvement ; we replace this with a — 2 D. cylinder, but patient is in doubt whether this is any better. We therefore give preference to the weaker cylinder. Our combination now is :

R. + 2 D. cyl. ax. 165° ◯ — 7 D. cyl. ax. 75°

with which vision equals $\frac{20}{80}$.

With the concave cylinder divided into two, it is almost impossible to rotate together in order to determine whether or not they are placed at the proper meridian. In fact we are unable to rotate the convex and the concave cylinder together, and under such circumstances it is customary to transpose to the equivalent spherocylinder, where there is only one cylinder to rotate. But in this case the transposition yields a — 9 D. cylinder, and unfortunately we do not have a cylinder of this strength in our trial case.

THE ONLY ALTERNATIVE

The only thing we can do is to take our strongest cylinder (— 6 D.) and find the best position for its axis, which the patient very confidently locates at 75°, and as this meridian is corroborated by the ophthalmometer, we have no hesitation in accepting it as correct.

The pin-hole disk showed the vision of the left eye to be capable of the greater improvement, and hence we commence the test of this eye with more hope. We try of course convex cylinders first, and the strongest accepted is + 1.50 D., with axis at 15°. We then use the concave cylinders with axes at right angles, increasing 1 D. at a time and securing greater improvement in vision, until we reach — 6 D. cylinder, axis 105°, with which combination vision equals $\frac{20}{30}$. We cannot transpose to a spherocylinder, and then verify the position of the axis, but we will have to use the — 6 D. cylinder alone, with which the axis is quickly located at 105°, which is verified by our ophthalmometric examination.

AGE A CONSIDERATION

This man is forty-six years of age, and some correction must be given for the presbyopia that is usually present at this age. He asks us if he cannot have his glasses in the bifocal form. We tell him these would not be satisfactory in view of the high degree of defect and the lowered acuteness of vision. Instead we will prescribe + 1.50 D. spheres in an extra front for reading.

This seems to be a favorable case for ordering the lenses ground in toric form, which would be as follows : the front surface of the lens ground with a + 5 D. sphere ; the posterior surface of the lens ground toric, — 12 D. at 105° and — 3 D. in the 75th meridian.

We are considering the right eye, and in order that you may follow me and understand this transposition, I will make two diagrams on the blackboard ·

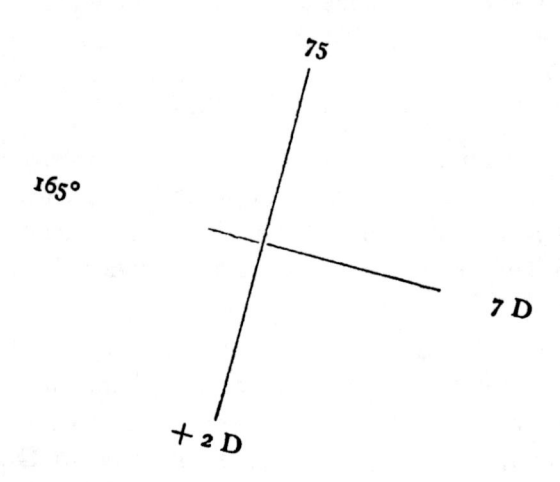

75

165°

7 D

+ 2 D

The cross-cylinder showed + 2 D. power in the seventy-fifth meridian and — 7 D. power in the one hundred and sixty-fifth meridian, and the toric lens or any other transposition must show the same power in the same meridians.

Now let us analyze the toric combination

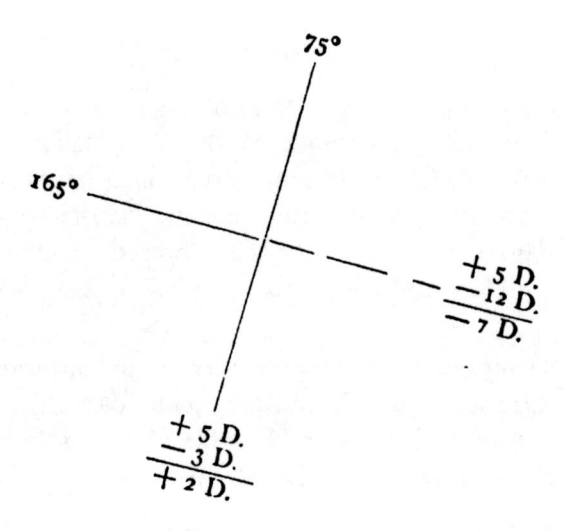

75°

165°

$$\frac{\begin{array}{r}+5\,D.\\-12\,D.\end{array}}{-7\,D.}$$

$$\frac{\begin{array}{r}+5\,D.\\-3\,D.\end{array}}{+2\,D.}$$

The + 5 D. sphere ground on the anterior surface of the lens affords + 5 D. power in both meridians. The — 3 D. ground in the seventy-fifth meridian reduces this meridian to + 2 D., and — 12 D. ground in the one hundred and sixty-fifth meridian changes this meridian to — 7 D., thus showing the same powers in each meridian as the original cross-cylinder. This toric lens will be of much better shape than the cross-cylinder or its equivalent spherocylinder.

As conical cornea is not very common, I am glad to be able to present this case for your study and observation. Except very rarely it is not a congenital disease, but makes its appearance about the tenth year. It is most frequently observed between the ages of fifteen and thirty. This gentleman tells us his vision was good and he did not commence to wear glasses until about fifteen years ago, which would make his age thirty-one when it was first noticed. The statement is made that women are more often affected than men.

AS TO THE CAUSE OF CONICAL CORNEA

The cause giving rise to this condition is not known, neither is the method by which it is produced, but it has been noted that a great many of the patients who have been affected were in feeble health. The integrity of the cornea is impaired so that it gives way at the point of greatest weakness, which is usually a little below its center, where it is forced forward and becomes sharply curved while around it the cornea assumes a conical form.

If the protrusion is moderate, the cornea maintains its transparency, or shows but slight opacity on oblique illumination. If the protrusion is extreme (and it is said to amount to as much as a half inch sometimes) the opacity is much more noticeable. Both eyes are usually affected ; the protrusion may slowly increase for a time, then become stationary and perhaps change again later in life.

Its most important effect is its influence on the acuteness of vision and the refraction of the eye, the cornea, on account of its increased curvature and displacement forward, renders the eye highly myopic and astigmatic. This impairs the vision so greatly that but seldom can lenses be found to raise vision to anything like the normal standard. The lenses giving best vision are usually strong concave spheres combined with cylinders. In this case, as you see, the strong lens is a concave cylinder combined with a weak convex, which is somewhat exceptional.

While on this subject of conical cornea, it may be interesting for you to know that many years ago Sir John Henschel proposed to correct the refraction by placing over the cornea a transparent shell or cup, which was called a "contact glass." The lens was ground in the shape of a meniscus, so that the posterior surface fits the front of the eyeball, somewhat after the nature of an artificial eye, while the front surface is ground to correct the refractive error. Unfortunately the contact glass acts as a foreign body, and on account of the irritation which it produces, cannot be long tolerated.

As optometrists, you are not so much concerned in the surgical aspect of conical cornea, but it will be interesting for you to have some knowledge of the operative treatment.

Sometimes an iridectomy is performed for the purpose of admitting light through a peripheral portion of the cornea, thus limiting the diffusion of the retinal image.

Attempts have also been made to flatten the cornea by excision of a small piece from the apex of the cone, or by touching and perforating the apex with the galvano-cautery, resulting in the production of a flattened cicatrix.

The extended use of myotics seems to act favorably in cases of conical cornea, the progress of which they moderate or check by reducing the tension of the anterior chamber. And, besides, they tend to improve the vision by contraction of the pupil.

The Value of Retinoscopy

[CLINICS No. 26]

I have to present before you to-day a little child, Dorothy B. K., five years of age. Her father tells us that about six weeks ago he noticed a turning in or convergence of the left eye.

You will remember I have always taught you that the first step in the examination of every case is to ascertain the acuteness of vision. Now here we have a child but five years of age, and her father tells us she does not know her letters. It is therefore a manifest impossibility in this case to determine acuity of vision, as this depends entirely on the answers of the patient as to the smallest letters he can see and name. Hence we will be compelled to approach and handle this case in a different way.

What is the first thought that occurs to you? It should be that this is a typical case of convergent strabismus dependent upon hypermetropia. In former times and even at the present day, many curious suggestions have been made by anxious parents as to the cause of squint. When Donders published his great work, his "Accommodation Theory," as the cause of convergent strabismus, was immediately and widely accepted.

ACCOMMODATION AND CONVERGENCE

I will refresh your minds with a brief mention of the principles involved. In the emmetropic eye the functions of accommodation and convergence are closely related, so that with every effort of accommodation there is a corresponding effort of convergence. This natural association of the two functions is disturbed in both hypermetropia and myopia.

A hypermetropic eye in a state of rest is out of focus for all objects, both near and far. Therefore, such an eye in order to see distinctly, must use its accommodation for distant vision to a degree corresponding to the amount of its hypermetropia. In near vision it must accommodate both for the hypermetropia and the nearness of object.

The excessive amount of accommodation required under such conditions causes a proportionately abnormal amount of convergence to be called into play. There was only enough accommodation used to focus the print, but there was more than enough convergence for binocular vision, and consequently one eye fixed the object and the other was compelled to deviate inwards.

It is a well-established axiom that the first step in rectifying any abnormality is to seek out and remove the cause. Therefore, if hypermetropia is the cause of convergent strabismus, a correction of the hypermetropia will cure the strabismus *if taken in hand early enough*, and herein lies the whole secret of the successful management of convergent strabismus by convex lenses.

In the large majority of cases the deviation appears very early —that is, in the fourth or fifth year, just about the time when the child begins to use its eyes in near vision, and that is the time when the glasses must be worn. If a case of convergent strabismus applies to you after the age of ten do not hold out any hopes of curing the condition by convex lenses. It may be necessary to order glasses for the correction of the existing hypermetropia, but the favorable period for restoring the eyes to parallelism has probably passed.

If we are correct in our diagnosis that this case is one of convergent strabismus caused by hypermetropia, the deviation will probably quickly yield to the restraining influence of the convex lenses because the child is coming under treatment at the age most favorable for satisfactory results.

OBJECTIVE METHOD OF EXAMINATION

Therefore, we must at once proceed to determine if hypermetropia is present, and if so to what extent. As the patient is too young to answer questions, or at least that dependence may be placed on her answers, the subjective method of examination is of little or no value. Instead we must depend upon information we gain ourselves by direct observation of the action of the eye, and this constitutes what is known as the objective method of examination. The condition of the refraction of an eye can thus be determined by the observation of the optometrist, and it follows as a matter of course that the correctness of the results depends largely upon the skill of the operator.

Within the past ten or fifteen years great progress has been made in the science of adapting lenses for the correction of refractive errors, and this advance is much more noticeable in the objective methods than in the subjective, the former being the more scientific of the two. And of all the objective methods, there is none more exact than retinoscopy, especially in the hands of an expert. Its value cannot be over-estimated in the young, in the feeble minded, the illiterate, and in cases of nystagmus, amblyopia and aphakia.

You must learn retinoscopy, both theoretically and practically. There may be some who can learn the theory and yet not be able to put it into practical use, but it is scarcely possible for any one to become a good practical operator without a knowledge of the theory. Both theory and practical application should be learned together, but a thorough knowledge of the former should always precede the latter.

THE RETINOSCOPE

I hold in my hand a retinoscope, which, as you see, is simply a mirror to project light into the eye, with an aperture for the eye of the observer, and without the disk of lenses you are accustomed to see in connection with the ophthalmoscope. While the mirror may be concave or plane, the latter is preferable.

Concerning the source of light, there is some difference of opinion as to the kind and condition of light and its position, but I feel like saying that too much importance has been attached to this matter.

As you see we use here a sixteen-candle power electric lamp with frosted globe and perforated asbestos chimney ; this is very convenient and always ready for use, but if electricity is not available, a Welsbach gas burner or an oil lamp of the student type would answer the purpose equally well.

The light may be on either side of the face or above the head, but it should be behind the plane of the eyes, and so placed that it may be easily and quickly projected into the eye of the patient by means of the retinoscope. I want to say right here that we are not so much concerned in the character of the *entering rays* as in the *emergent rays*, which vary in accordance with the refraction of the eye.

Light emerging from an emmetropic eye the rays will be parallel.

Light emerging from a hypermetropic eye, in which there is a deficiency of refractive power, the rays will be divergent.

Light emerging from a myopic eye, in which there is an excess of refractive power, the rays will be convergent and will come to a focus at some definite distance at what is called the far point, which can be located by the optometrist in the practice of retinoscopy.

If the eye is not naturally myopic, it is made so artificially by interposing certain lenses, by which the far point is located at some definite distance, which is called the working distance. By common consent this is placed at one meter.

This child is so small that we will ask her to stand up instead of allowing her to be seated, as we usually do, and as I sit down her eyes and mine are nearly on the same level. We face each other squarely and I ask her to look not at me nor at the light, but past my head at the opposite side of the room.

It is not necessary that the room be made dark as night, but I will ask that the shades be drawn down, in order to darken the room somewhat, and that there may be no other source of light in the room during the test except that used in connection with the instrument. This favors a dilatation of the pupil, so that we may obtain a good reflex.

USE OF THE RETINOSCOPE

I take the retinoscope in my right hand and rest it against the side of my nose and my forehead and directly in front of my right eye.

It might be well to mention to you that if you have any error of refraction in your own eye of such a degree as to impair your visual acuity, it is desirable that you wear your own correcting lenses in using the retinoscope, because you want the best possible vision in studying the action of the reflex. This may prove somewhat annoying at first, owing to the reflections from the surfaces of your lenses, but a little experience will enable you to overcome this. You will understand that your accommodation is not taken into consideration, but simply that your visual acuity may be good enough to enable you to recognize what you see.

I now project the light on the face, and I must say to you that to properly control this beam of light requires considerable practice. When you see some one else doing it, it seems like a simple matter to hold the mirror so that the light will reach the eye of the patient; but I wish to warn you that on your first trial of retinoscopy you will find some difficulty in locating the eye.

As I project the light upon the eye, you will see a circular spot of light upon the face surrounding the eye, which is sometimes called the "light area." As the retinoscope is tilted this light area is made to move, and it always moves on the face in the same direction as the mirror is tilted; a very slight movement sufficing to cause the light to pass across the face of the patient.

HANDLING THE MIRROR

Some authorities advise that the retinoscope be held firmly against the forehead, and the desired tilting of the mirror be produced by a movement of the head, to the right and left, or up and down. This method does not appeal to me, as the movements of the optometrist's head give the appearance of awkwardness. But, instead, the mirror should be rotated or tilted by a movement of the wrist alone, the instrument being held in a vertical position to get the side-to-side movements, and in a horizontal position to obtain the up-and-down movements.

When the light is reflected into the eye of the patient, the pupil loses its darkness and becomes luminous, and the bright reflection we see is known as the retinal reflex. Those of you who are close to me can see the light area on the face, and can see it move as I tilt the mirror; but you cannot see the retinal reflex in the pupil. Only the eye behind the sight hole of the retinoscope perceives this, because the emergent rays from the eye under observation pass back to the source of light (the mirror) and enter the eye of the observer through the sight hole.

Now, while the movement of the light area upon the face is always in the same direction as the movement of the mirror, the movement of the reflex may vary, being with or against the movement of the mirror, according as the emergent rays are divergent or convergent, and it is this movement of the reflex that determines the refraction of the eye.

With a plane mirror retinoscope at a distance of one meter, we obtain the following indications, which I will mark on the blackboard for your guidance·

No movement of reflex — myopia of 1 D.

Movement of reflex "with" { Emmetropia, Hypermetropia, Myopia less than 1 D.

Movement of reflex "against" — myopia over 1 D.

The several steps which I follow in this little girl's case are as follows:

1. Illuminate the pupil by directing the light full upon the eye, so that the pupil is in the center of the light area.

2. Rotate or tilt the mirror, a slow movement being preferable, in order to give the mind time to interpret what the eye sees.

3. Note whether the reflex is stationary or movable, and, if the latter, whether with or against.

If there is a movement of the reflex, find lens that neutralizes the same.

In this case, as I slowly rotate the retinoscope to my left, I see the light area on her face and the reflex in her pupils both move to my left; in other words, the movement of the reflex is "with" the light area. By a reference to the table on the blackboard, you will see that the condition of the refraction in this case is either emmetropic, hypermetropic or myopic less than 1 D. Now we must use convex lenses to determine which of these conditions is present.

I use + 1 D. lenses first, placing them in the trial frame in front of her eyes. I again rotate the mirror and cause the light to travel across the pupils, and find the movement is still "with." Now what have we learned?

We know that the eyes are hypermetropic, and we have by this + 1 D. lens eliminated both emmetropia and myopia. This lens "allows" for the distance at which the test is made, and it is the custom of many skiascopists to commence the test at once with this lens before the patient's eyes, in order to save time, in which case no movement indicates emmetropia, with movement hypermetropia, against movement myopia.

Knowing, then, that we have here a case of hypermetropia, we must increase the strength of the convex lenses until we neutralize the movement. We replace the + 1 D. spheres with + 2 D. and find the movement is still with. We now try + 3 D. and find the movement is still in the same direction, but with a + 4 D. the movement is neutralized. In order to make sure, we try + 4.50 D., with which the movement is against, showing over-correction. We then try + 3.50 D., with which the movement is with, showing under-correction.

Although + 4 D. lenses neutralize the movements of the reflex, don't jump too hastily to the conclusion that they represent the degree of defect, because we must first make allowance for the working distance, and this is done by adding — 1 D. to the findings of the retinoscope. This addition is made in all cases, whether the neutralizing lenses be convex or concave.

PRESCRIBING GLASSES

In this case we have the following problem, which I will place upon the blackboard:

$$+ 4 \text{ D. S.} \quad \text{Retinoscopic finding.}$$
$$\text{Add} - 1 \text{ D. S.} \quad \text{Allowance for working distance.}$$
$$+ 3 \text{ D. S.} \quad \text{Amount of hypermetropia.}$$

This is algebraic addition, but some students prefer to fix the rule in their minds as follows: In hypermetropia subtract 1 D. from the retinoscopic findings, and in myopia add 1 D.

Having thus measured the amount of hypermetropia in this case, it remains to prescribe such lenses as will afford comfort and satisfaction, and at the same time check the tendency to excessive convergence. There is room for difference of opinion on this point; I do not think it wise to attempt to force the eye to accept the full correction at once, and especially in a child who has never worn glasses. We are usually guided somewhat by the effect of the convex lenses on distant vision, which would be impaired by the full correction, and this would be a manifest disadvantage in school work. As this patient is unable to give us any information on this point, we will venture to correct two-thirds of the defect, and will order $+ 2$ D. for constant wear, and will emphasize our instructions to the father that Dorothy should not be allowed to remove her glasses during waking hours. This is not so much on account of the hypermetropia, but because it is complicated with strabismus.

A Case of Pigmentary Retinitis, Illustrating the Value of Ophthalmoscopy

[CLINIC No. 27]

A. E. S., aged fifty-two years, wood-worker by occupation, German by birth. This is the name, age, occupation and nationality entered on the card of the patient who stands before us. He complains that his vision is poor and especially so at night.

After some questioning we are able to get a history of this case somewhat as follows: Vision has been fairly good during the day, but very poor at night except in a very bright light. He has, therefore, been accustomed to remaining indoors at night, and occasionally when he did go out in the evening he never left the house except in company with some one on whom he could rely. In the short days of winter he hurries home before twilight, and he has learned to dread dark and rainy days. He came to this country when a boy, and he remembers he had this night blindness at that time, but he feels that his condition has been gradually getting worse. He has been fitted with glasses a number of times but with little satisfaction. He is a single man, as he was advised by physicians never to marry, as his eye disease would probably be transmitted to his offspring.

We find his acuteness of vision to be $\frac{20}{100}$, the same in each eye. We make use of the ophthalmometer, which shows astigmatism against the rule, the excess of curvature in the horizontal meridian being about 2 D. This gives us a pointer as to the condition of the refraction, and we will proceed with the trial case to work it out.

If convex cylinders are required, the position of the axis would be at 180°, but a $+$ 1 D. cyl. placed in this position is rejected. A rotation of the cylinder to some other meridian makes vision still worse. We then try a concave cylinder axis at 90°, and this is at once accepted. This shows we are on the right track and we increase the cylinder and after a few trials we find the best vision is obtained with a $-$ 2 D. cyl. axis 90°, with which V. = $\frac{20}{50}$. The test of the right eye yields the same result.

Our failure to raise vision to normal by any combination of lenses, and the history of poor vision given us by the patient, leads us to suspect some diseased condition, and in order to determine this point we must make use of the ophthalmoscope.

INVENTION OF THE OPHTHALMOSCOPE

Prior to the invention of the ophthalmoscope, the interior of the eye was an unsealed book. The ordinary black appearance of the pupil was supposed to be due to absorption of the light by the pigment cells of the choroid. Doubt was first thrown on this idea about a hundred years ago; various observers worked upon this problem, but it was not solved until 1851, when Helmholtz contrived an eye mirror, by means of which he was enabled to illuminate the eye and obtain a view of the fundus.

The announcement was made that all eyes are luminous, but that the eye of the observer must be placed in the path of the returning rays. This is accomplished by means of a mirror with a perforation in it. When light is reflected from the mirror into the eye, it returns again to the mirror and part of it passes through the perforation into the eye of the observer placed behind it.

The examination of the fundus by means of the ophthalmoscope affords important information not only in intra-ocular diseases, but also in diseases of other organs, as the brain and kidneys, and often enables us to detect the presence of syphilis. Ophthalmoscopy to be of value requires some skill, not only to obtain a good view of the fundus, but also to correctly interpret what is seen, and it increases in difficulty as we pass to the observation and interpretation of all the finer changes produced by disease.

The purpose of the ophthalmoscope is to obtain a good view of the interior of the eye, and therefore in the selection of the ophthalmoscope, I may say to you that it is not at all necessary to purchase the most expensive and most complicated instrument, but I can assure you that the small ophthalmoscope, such as I hold in my hand, will answer every purpose.

EVOLUTION OF THE OPHTHALMOSCOPE

It is well for you to have some knowledge of the history of our science, and therefore it will be interesting for you to know that the original Helmholtz instrument consisted of three thin plates of glass, which formed the hypothenuse of a triangular box, which was closed on all the other sides and blackened in the interior. The small side of the triangle contained an eye-piece for the use of the observer.

At the present time this instrument is little more than a curiosity, having given way to the improved ophthalmoscopes of the present day, and yet by reason of its weak illuminating power, it is still of value in the detection of fine opacities in the vitreous.

The most popular ophthalmoscope is the Loring. The mirror is concave, having a radius of curvature of sixteen inches and a focal distance of eight inches. It is oblong in shape, and can be tilted to an angle of twenty-five degrees. A concave mirror is preferable because it gives a stronger illumination.

Considerable stress has been laid upon a "dark room," by which we understand a small room whose walls and ceiling have been covered in black. While this may be desirable, it is not at all essential; by simply drawing the shades, you can convert your office into a dark room, as occasion requires, that will answer every purpose. As you gain in experience extreme darkness becomes less necessary, and an expert ophthalmoscopist often makes his examination without taking the trouble to draw down his shades.

For illumination, if electricity is available, it is certainly the most convenient. A frosted glass should be used to avoid the annoying reflection of the wires in the ordinary lamp. Gas is perhaps the most common and convenient illuminant, and may be used in connection with a Welsbach burner, which gives a beautiful illumination, and is preferred by some observers to electricity.

If gas or electricity is not available, an argand oil lamp will be found to answer the purpose. If the luminous ophthalmoscope is made use of, the question of light is settled at once, as this instrument carries its own lamp.

Let me refer again to the mirror, which as I told you is concave and has a focal distance of eight inches. Parallel rays falling upon such a mirror would be reflected to a point at a distance of eight inches; and conversely, if the light is placed eight inches from the mirror, the rays of light would be reflected parallel. If the light is nearer than eight inches, the rays are reflected divergently.

In most cases we make use of convergent rays, which are obtained when the light is placed farther than eight inches, the degree of convergence increasing with the distance of the light.

THE USE OF MYDRIATICS

In the various works on the ophthalmoscope, which are written by medical men for medical men, constant reference is made to the use of mydriatics. But this need not deter you in the great majority of cases; after a little practice you will be able to make an ophthalmoscopic examination with only the dilatation of pupil produced by the darkened room.

The optic disk is the landmark first looked for at the fundus of the eye, and fortunately it is not so acutely sensitive, which allows the light to be turned in this direction without inconvenience, and hence this part can be most easily examined without a mydriatic. This is in contrast to the sensitiveness of the region of the yellow spot, as shown by the extreme contraction of the pupil when light is thrown in this direction; in addition to which there are also corneal reflections which often interfere with the view.

We will examine this man's eyes first by the direct method, and if you will watch me closely, you will see the various steps necessary. We begin our examination with the ophthalmoscope held at a distance of twelve or fifteen inches; we reflect the light into the eye and observe the red pupil reflex. The brightness of this reflex may be modified by the size of the pupil, the transparency of the media, the refraction of the eye, and the amount of pigment present. In this case I find the reflex not nearly so bright as from a healthy eye.

I now approach the patient while still holding the ophthalmoscope in front of my eye, and keeping the pupil brightly illuminated. The beginner will find some difficulty in maintaining the red reflex as he changes his position. But this can be learned by a little practice, and approaching slowly I get my instrument as close to the patient's eye as spectacles are worn. The direct method is so called from the fact that we look directly into the eye.

As we look into an eye with the ophthalmoscope, what do we see? The normal eye ground is reddish on account of the vascularity of the choroid showing through the transparent retina. The eye ground of a blonde would be pinkish, while that of a brunette would be much darker, on account of the difference in the amount of pigment in each. The usual color of the eye ground is sometimes described as "orange-red."

EXAMINATION OF THE OPTIC DISK

We now look for the optic disk (sometimes called simply the disk), which is the intra-ocular ending of the optic nerve. The actual size of the disk is 1.5 mm., which as most of you know is about $\frac{1}{16}$ of an inch. Its position in the fundus is about 10° or 12° to the inner side of the posterior pole of the eye. It corresponds to the normal blind spot in the field of vision.

The shape of the disk varies; it may be circular or oval, or perhaps irregular in outline. In most cases it is slightly oval vertically. When it appears decidedly oval, we suspect astigmatism. In health the margin of the disk is distinct and well defined.

Near the center of the disk there may frequently be seen a slight depression, and as it is not abnormal, it has been termed the 'physiologic cup.'' It is caused by a separation of the nerve fibers after passing through the lamina cribrosa.

At the bottom of the cup there is frequently seen an area composed of little gray spots (representing the optic nerve fibers), with white interspaces (representing the lamina cribrosa), which is the sclerotic through which the nerve fibers enter.

Encircling the disk may be seen the scleral ring, which is light colored, or the choroidal ring, which is pigmented. Either of these may appear in the form of a crescent instead of a ring, and they may be present or absent.

The optic disk is whitish in color and at its center the blood vessels enter and radiate over the retina. The veins are recognized as being larger and darker. The retina being transparent is almost invisible, but it is located by the retinal vessels.

The macula lutea, or yellow spot, is about 3 mm. to the temporal side of the disk. It is oval horizontally, and is darker in color than the rest of the fundus. At its center is the fovea centralis, which is very small and appears as a bright spot.

WHAT THE OPHTHALMOSCOPE REVEALS

I am now in position for direct ophthalmoscopic examination, and as I look into the eye I see patches of pigment scattered all over the retina, which at once leads me to make a diagnosis in this case of retinitis pigmentosa, or pigmentary degeneration of the retina. The changes in this instance are more degenerative than inflammatory, and usually, as in this case, both eyes are affected.

I find the optic disk in this case yellow in color, and the retinal vessels few in number and small in size.

The pigmentation occurs in patches and is deposited in the retina along the course of the blood vessels. In the early stages of the disease, the characteristic pigment spots are at first seen only at the periphery and are few in number; but as the disease progresses the spots increase in number and become connected together by branching threads. This irregular network of pigment slowly creeps towards the disk and at last invades the yellow spot. In the case of this man, in whom the disease was probably congenital, the whole fundus is pretty well covered.

This disease is also marked by diminution in the number and size of the retinal vessels, sometimes contracted to mere threads.

The further progress of the degenerative changes is shown by the grayer and more leaden hue of the fundus, by the closer and more marked network of the pigment, by the increased atrophy of the disk, and by the further diminution in the size of the retinal vessels. There is also great contraction in the field of vision.

Usually, the diagnosis of a case of this kind is not difficult. In the early stages when the deposit of pigment is scanty and confined to the periphery, the disease may be overlooked. But a little later on, when the disease has extended to the more central parts of the field, it cannot be mistaken.

Sometimes the disease may remain at a standstill for years, but usually the prognosis is unfavorable, because the changes progress until at sixty years or more all serviceable vision is lost.

PIGMENTARY RETINITIS

Let me briefly enumerate the chief subjective symptoms of pigmentary retinitis. The first to attract the attention of the patient, and the chief characteristic symptom of the disease, is night blindness. He finds his vision is decidedly poorer as soon as twilight begins and after the sun has gone down. He may see well in daytime, but after dark he stumbles and bumps into objects.

After a time his vision becomes impaired at all times, and patient appears to be near-sighted from the fact of holding objects close in order to see them better.

The field of vision becomes markedly contracted, so much so that patient is unable to go out alone.

In advanced cases, nystagmus may set in, which is shown by a rapid lateral oscillation of the eyeballs.

I am glad to have the opportunity to present before you today this case of retinitis pigmentosa, which is almost typical, both in the subjective symptoms and in the ophthalmoscopic appearances. I hope this case will make impression on your minds, so that you will not fail to recognize the disease in after years if it should fall under your notice.

This is distinctly a diseased condition, and therefore even though you are able to raise the acuteness of vision by glasses, I would advise you not to prescribe them, but rather to explain the condition to the patient or his family, and shift the responsibility of such a serious condition to a medical man.

Albuminuric Retinitis, or the Retinitis of Bright's Disease

[CLINIC No. 28]

H. W. H., aged forty years, complains of foggy vision and that he has been unable to get glasses that were satisfactory. First noticed the impairment of vision about six months ago.

We take the acuteness of vision and find it to be : O. D. $\frac{20}{200}$, O. S. $\frac{20}{100}$.

We turn to the ophthalmometer, the readings of which are 50 D. excess in the horizontal meridian, showing the existence of astigmatism and classing it against the rule.

We return to the test case and after a few trials with convex spheres and cylinders, we decide on the following formula as the best we can do :

$$+ .50 \text{ D. sph.} \bigcirc + 1 \text{ D. cyl. axis } 180°$$

The same for each eye, raising the vision of right eye to $\frac{20}{100}$ and of left eye $\frac{20}{70}$.

This result is unsatisfactory both for us and the patient. We will try the pin-hole disk in order to ascertain the possibility of further improvement in vision, but the result is negative, as the patient says he cannot see as well through the pin hole as with the glasses above-mentioned.

A DISEASED CONDITION INDICATED

This last test shows the presence of some diseased condition that is destroying the vision and interfering with the action of glasses. Our only solution of the case lies in the use of the ophthalmoscope.

We try the direct method, and as we approach closer looking through the illuminated pupil, we see at once that some abnormal condition is present.

As customary, we look first for the optic disk, which appears clouded and congested and somewhat swollen. We also see numerous white spots and patches of various shapes and sizes. The larger spots surround the disk, while in the region of the yellow spot they appear as fine white dots and lines, arranged in the shape of a star around the macula, or like the spokes of a wheel with the fovea as the center.

We also see extravasations of blood scattered here and there over the retina, as the result of recent hemorrhages.

We have a diseased condition here sure enough, and we have no trouble in locating it in the retina. The only question is as to the form of retinitis, and from my experience in similar cases I have no difficulty in diagnosing the case as one of albuminuric retinitis.

Most cases of retinitis are simply incidental symptoms of general constitutional disease, and albuminuric retinitis is one of them. This is a disease that cannot be modified by glasses, or even any local treatment of the eye, but on account of its seriousness the interest of the optometrist lies largely in its early diagnosis.

SYMPTOMATIC OF BRIGHT'S DISEASE

It may be stated as a general axiom, that retinitis occurs most often in connection with albuminuria, and hence the ophthalmoscope becomes of the greatest assistance in the diagnosis of inflammation of the kidneys, because albuminuric retinitis is one of the earliest symptoms of Bright's disease.

Failure of vision of recent occurrence and without apparent cause, should lead you at once to make an ophthalmoscopic examination, when the presence of retinal disease would be quickly detected, and if of the albuminuric variety you should advise patient to consult their family physician for an examination of urine, in order that your diagnosis may be corroborated. It has been estimated that one-fourth to one-third of all cases of kidney disease, show this form of retinitis. It may also occur in the nephritis that accompanies pregnancy, and that following scarlet fever.

Chronic kidney disease is a very serious malady that comes on gradually and insidiously. When in an advanced stage, it is beyond the reach of medical skill; although much may even then be done to prolong life and relieve suffering. But in its early stages it can frequently be arrested and sometimes radically cured. Therefore, any aid to an early diagnosis becomes of the greatest importance, and no warning eye symptoms should be misunderstood or disregarded.

For these reasons the fact should be known to every optometrist that there are certain changes in the retina that are pathognomonic of disease of the kidneys, and I would be failing in my duty if I did not emphasize this statement, so that it may be indelibly impressed upon the mind of every member of this class.

Usually both eyes are affected, but varying in degree. The ophthalmoscopic appearances bear no definite relation to the degree of impairment of vision, for preservation of useful sight is not incompatible with pronounced changes in the retina.

SIGNIFICANCE OF ALBUMINURIC RETINITIS

The detection of retinitis is not only of diagnostic value, but also of prognostic importance in affections of the kidneys. It has been stated upon good authority that life is rarely prolonged more than one, or, at most, more than two years, after the development of retinitis.

When retinal hemorrhages are present, the prognosis is much more unfavorable. Hemorrhage into the vitreous and detachment of the retina may also occur, but these conditions are not common.

Albuminuric retinitis also occurs in pregnancy, having been observed as early as the third month, although it is more frequent about the seventh or eighth month. As this is a serious condition for a pregnant woman and as vision is not always impaired early in the disease, systematic ophthalmoscopic examinations should be made during the course of pregnancy whenever albuminuria exists.

Authorities agree that the visual disturbances which make their appearance during the last weeks of pregnancy, although accompanied by the same retinal changes, are of less grave import in so far as sight is concerned, because in the great majority of such cases, vision becomes normal after labor.

Albuminuric retinitis is rarely monocular; both eyes are usually affected, but not always commencing at the same time or to the same degree.

The prognosis as regards the vision of patient and his life are both bad. Vision fails gradually at first, and in many cases this is the first symptom of which patient takes notice and which draws his attention to his eyes and through them to the condition of his kidneys.

The duration of life after the retina has become affected, is but little more than a year in the majority of cases, although, of course, to this statement there are many exceptions. Cases have been known to live three, five and even eight years, after the onset of the retinitis, when due to specific cause, by active and persistent treatment.

Bright's disease may cause a retinitis which does not present exactly the characteristic features of the disease as I have mentioned them to you.

In some cases hemorrhages may be the most conspicuous feature scattered over the fundus, and appearing as early manifestations before the disk has become involved. When the hemorrhages are absorbed the spot becomes white, and if the underlying choroid has been involved, these areas may be partially pigmented.

In other cases, it is the inflammation of the optic disk, which is swollen and congested, presenting the so-called "choked disk," or papillitis, as it is sometimes called. The edges of the disk are hazy and woolly.

In other cases the only changes noticed are a few fine spots about the macular region and one or more small hemorrhages. Hence in any case where the ophthalmoscope shows signs of retinitis, the patient should be advised to have careful and repeated examinations of the urine made.

A CASE FOR A PHYSICIAN

Albuminuric retinitis is most likely to occur among persons of middle age, who may apply to you for glasses for impaired vision, and it behooves you as educated optometrists to be able to recognize the presence of this insidious and dangerous disease, in order that you may be able to give the proper advice and thus reflect credit on yourself and your profession, while giving the patient the advantages of early treatment; instead of attempting to fit glasses for a diseased condition, thus losing time for the patient, and bringing yourself and your profession into merited disrepute.

The onset of the symptoms is often sudden, corresponding to the occurrence of a hemorrhage or swelling in the region of the macula. It is most likely that there are evident and characteristic changes in the fundus, consisting of alterations in the size, course and texture of the blood vessels, which precede the usual symptoms, but in these early stages there is little to attract the patient's attention to his eyes, and ordinarily relief is not sought until grosser alterations have taken place so that vision has become perceptibly impaired.

The pathological changes in the eye are primarily in the blood vessels and consist in a degeneration of their walls, and involving the vessels not only of the retina but also those of the choroid and other parts of the eyeball. The whitish spots seen with the ophthalmoscope are usually granular or fatty tissue degenerative products. The supporting fibres of the retina become swollen, infiltrated and also degenerated. There is serous effusion of the retinal structures. The nerve fibres are swollen and contain degenerative products such as granular or fatty particles.

It seems probable that the cause of all these changes is fundamentally an alteration in the character of the blood supplied to these tissues, but the precise nature of the toxic substances producing them has not yet been determined.

During the course of Bright's disease, periods occur when the amount of urea in the urine is considerably diminished. At such times there is usually an aggravation of all the symptoms, both ocular and general.

Whether the retinal disease is due to the retention of urea or of other excretory products in the blood is an open question, but it has been observed that an increase of urea products in the urine is usually followed by an improvement in the retinal condition. This does not prove conclusively that the urea is the real toxic factor in the production of the vascular changes in the retina during nephritis, but it at least serves as an index of the degree of toxaemia.

CHARACTERISTIC FEATURES

In summing up let me emphasize the characteristic features of albuminuric retinitis as revealed by the ophthalmoscope.

1. White spots and patches. These are well defined, of various shapes and sizes, and occur in the region of the macula and optic disk. In typical cases they are arranged in lines radiating from the center of the macula, sometimes in all directions, sometimes only in a certain portion of the field. Other and larger spots may be seen in the area surrounding the disk. These may coalesce and form a ring-shaped zone around the disk.

The star-shaped appearance of the lines radiating from the macula, was at one time supposed to be absolutely characteristic of Bright's disease, but too much dependence must not be placed upon it in the diagnosis, as it sometimes fails to appear in actual albuminuric retinitis, while at other times it is found in other forms of retinitis.

The white massings around the optic disk have been referred to by some authorities as "snow banks," and are somewhat peculiar to albuminuric retinitis, as in other forms of retinitis, the exudate is not so rounded but elongated, and follows the course of the large vessels.

2. Hemorrhages. Here and there are retinal hemorrhages, occurring at any time during the progress of the disease. They may be rounded or flame-shaped, occurring in the nerve fibre layer and in the neighborhood of the vessels. They may appear as mere flecks of blood or large extravasations, according to the intensity of the disease. Often old hemorrhages in the shape of pigment spots or white patches, may be seen along with fresh extravasations. Small hemorrhages in the deep layers, when occurring early in the disease, are considered as unfavorable symptoms. The blood may disappear by absorption, without leaving any white or pigmentary remains.

3. The optic disk may be unchanged, but it is more likely to show evidence of congestion or inflammation, presenting the appearance of a neuritis, varying from a slight redness with blurring of its outlines to extreme swelling as in the choked disk of brain tumor. When the disk is much affected, the retinal lesions are generally more extensive than are found in connection with neuritis from other causes.

4. The appearance of the retinal vessels. The veins are distended and often tortuous. The arteries may be normal or sometimes narrowed, not infrequently showing whitish streaks along one or both borders, which in some places may obscure the vessel itself, and convert it into a white cord. The vessels may be partially buried in the swollen retina, and are seen to cross the white patches which lie in the outer layers of the retina, and are concealed by those patches which lie in the nerve fibre layer.

The ophthalmoscopic appearances may change slowly. The whitish spots may gradually increase in number and may coalesce into large patches. New hemorrhages may appear and old ones vanish. In general, except in the few cases that recover, the ophthalmoscopic picture becomes more and more complicated.

RECOMMEND PATIENT TO PHYSICIAN

After having made your diagnosis in a case of this kind, you must refer your patient to a medical man and wash your hands of any further responsibility.

In regard to treatment, I may say for your general information, that the eye should have rest, as in any other form of retinitis.

Errors of refraction or failure of accommodation should be carefully met by proper lenses, to be worn whenever the eyes are used, which must be little. Smoke glasses are often ordered.

Beyond this, the treatment is that of the general condition · regulated diet, woolen clothing, a dry, mild climate, avoidance of worry, and medicines directed to the renal condition.

The subject of treatment offers but little of comfort either to patient or physician in chronic, well-marked cases. Still it must be noted that periods of improvement are often observed in the course of the disease, and that the retinitis of acute Bright's disease or of pregnancy, may entirely clear up with the improvement or cure of the nephritis.

Accommodative Esophoria

[CLINIC NO. 29]

This young lady, Miss Nanna McK., is twenty-seven years of age. She complains of pain in eyes and headache, and says she can't see well.

We find the acuteness of vision in each eye to be $\frac{20}{20}$ partly. On examination of near vision, we find the near-point to be six inches. This recession of near-point at once indicates the presence of hypermetropia. You will probably recall from your study of the amplitude of accommodation at the various ages, that the near point at this age should not be farther than five inches.

We turn to the ophthalmometer, which gives no evidence of astigmatism beyond the normal amount of slight excess in the vertical meridian of the cornea.

As the vision is so nearly normal, we can exclude myopia, and as the ophthalmometer shows the meridians of the cornea to bear the normal relation to each other, we may exclude astigmatism. The diminished amplitude of accommodation points to hypermetropia, for the detection and measurement of which we will now direct our efforts. In cases like this where the patient is able to name some or all of the letters on the No. 20 line, we cannot expect to afford much improvement in vision by convex lenses, because any hypermetropia that may be present would exist in the form of latent rather than manifest hypermetropia.

METHOD OF EXAMINATION

Therefore, we will proceed with the fogging system, with the details of which you are by this time more or less familiar. We try the right eye first, placing before it a $+ 5$ D. lens. This fogs vision to the extent of blotting out the whole card except that the No. 200 letter can be guessed at. A $- .50$ D. placed before it improves vision, and we increase the concave lens until $- 2.50$ D. is reached when the No. 20 has now become legible. The difference between the two lenses, or more strictly speaking the algebraic addition of the lenses, shows the amount of hypermetropia we have been able to uncover, viz., 2.50 D. We repeat the test with the left eye and obtain the same result.

We now make use of the Maddox rod, placing it over the left eye in a horizontal position, and directing the patient's attention to the small point of light across the room.

We ask the patient on which side of the light the red streak appears, and she replies that she doesn't see any red streak. This is not at all unusual because the image of the uncovered eye is so much brighter that it entirely occupies the attention of the brain to the exclusion of the distorted image of the other eye.

But an inexperienced man must not get discouraged and jump to the conclusion that he is therefore unable to test the muscle balance. We simply rotate the Maddox rod and the attention of the brain is at once called to the moving object and the red streak caused by the rod becomes visible.

Now, in answer to our inquiry, she is able to locate the red streak and she tells us it is away off to the left, perhaps a foot or more from the light. Now, then, what have we? We have produced an artificial diplopia by means of the Maddox rod, and the image of the left eye (the red streak) is seen to the left, and the image of the right eye (the natural light) is seen to the right. This form of diplopia, which we have produced, is classed as *homonymous*, and is due to an excess of convergence, and therefore, we have here a case of esophoria.

How do we measure the amount of the esophoria? By the strength of prism base out that is necessary to bring the red streak back to the light. We try a 5° prism; this brings it closer, but it is still to the left. An 8° prism brings it still closer, and a 10° prism base out causes the streak to pass vertically through the flame, and is, therefore, the measure of the esophoria.

In spite of the large amount of esophoria, we have in this case single binocular vision, because of the desire on the part of nature for single vision. If the need for this desire is removed by excluding one eye from vision, as I do in this case while the patient looks at the letters with the other eye, I can see the covered eye deviate inward. As I remove the cover, the eye quickly resumes its proper position. Such a condition has been termed *latent* strabismus.

LATENT STRABISMUS

When the inward tendency is very strong, an excessive strain is imposed upon the nerve centers to supply sufficient innervation to the external recti muscles to counterbalance this tendency to extreme convergence. Beyond a certain point this effort cannot be maintained, and vision is then performed by one eye, while the other eye deviates inward, or, in other words, assumes its position of equilibrium. This constitutes *manifest* strabismus.

Latent strabismus may become manifest at certain times, as when the eyes are tired from prolonged use, and especially in near vision, when spasm of convergence may be excited under conditions similar to those that produce spasm of accommodation. Under normal conditions there is call for accommodation and convergence in equal proportion. At ten inches there is used 4 D. of accommodation and 4 M. A. of convergence. The association between the functions of accommodation and convergence is so intimate that exercise of one is voluntarily accompanied by a corresponding action of the other.

In spite of this, the connection between the two functions is not so strong but that each may suffer variation within certain limits, and thus render distinct binocular vision possible in ametropia, where one or the other function needs to be used in excess of the other.

In hypermetropia excessive convergence is provoked, because the inordinately great accommodative effort required to overcome the diminished refraction and maintain distinct vision gives rise to the impulse for more convergence than the distance of the object demands. This excess of convergence may be latent as in esophoria, or it may be manifest as in convergent strabismus, but in either case the common cause is hypermetropia.

When we release the eyes from the necessity for binocular fixation, and allow them to assume their positions of equilibrium, the inward deviation manifests itself by the appearance of the diplopia that has been produced. This is accomplished by making one retinal image so dissimilar from the other that there is no desire to fuse them into one. This we have done in the case before us; and you will remember the red streak seen by the left eye appears to be way off to the left. Inasmuch as the eye turns toward the right, it would seem on first thought that the object seen by this eye should appear to the right, instead of the left, as it actually does.

HOMONYMOUS DIPLOPIA OF ESOPHORIA

As this seeming contradiction is very confusing to optical students, I think we can spend a few minutes profitably in discussing it. I can possibly make the matter clearer to you by a diagram on the blackboard·

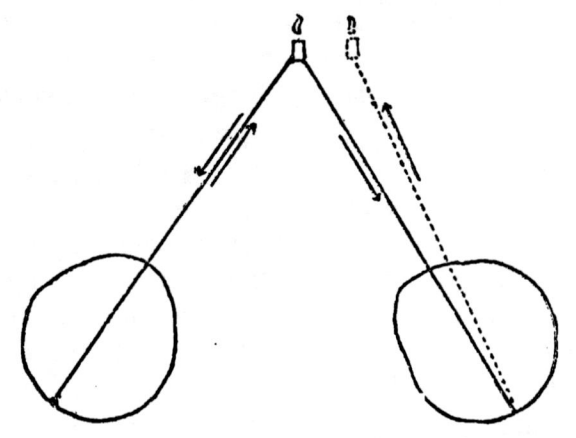

Diagram illustrating the homonymous diplopia of esophoria

In this diagram you will see that the light from the candle falls upon the yellow spot of the left eye, forms an image there and is referred back in the direction from which it came. The rays from the same candle entering the right eye do not fall upon the yellow spot but strike the retina at the inner side of it. Now, then, according to the law of projection, as you learned it while studying the physiology of vision, the light is referred, not in the direction from which it actually comes, but in the direction from which it appears to come, and thus is seen to the right. An object situated to the right impresses its image on the left of the retina, and is referred by the brain from left to right. An object situated to the left impresses its image on the right of the retina, and is referred by the brain from right to left.

In like manner when an impression is made upon the upper part of the retina, it is referred by the brain to the lower part of the field where the object lies from which the impression is received. When an impression is made on the lower part of the retina, it is referred upward.

And so in the case under consideration where the impression is made on the retina to the inner side of the yellow spot, it is referred outward. This is the law of projection, and it is by this law that our vision is erect in spite of the fact that the retinal image is inverted. I trust that you will all now understand how and why the false or second image is seen in the opposite direction from the deviation, and that therefore the diplopia of esophoria is homonymous.

The symptoms of esophoria are not distinctive; slight degrees may give rise to no discomfort whatever. The least amount which is likely to cause asthenopia cannot be definitely stated, since this will vary with the nervous susceptibility of the individual patient. In general, it may be said that an esophoria of 2° or 3° is not much beyond the limits of normal muscular equilibrium, when tested at a distance of twenty feet.

Headache is a symptom of esophoria, coming on periodically and sometimes accompanied with vertigo and nausea. It occurs after the use of the eyes for distant vision more often than near vision, and is due not so much to extra effort of the internal recti as to the strain placed upon the external recti to prevent undue convergence, as otherwise diplopia is likely to result. This may afford relief to the headaches but the patient will be greatly annoyed by the double vision. The strabismus has now become manifest and the asthenopia is replaced by a new train of symptoms originating from the loss of binocular vision.

But the diplopia does not continue; monocular vision is soon established, as a result of the cultivated habit of disregarding the false image, which is greatly favored by the insensitiveness of that part of the retina upon which the false image is impressed.

TREATMENT OF ESOPHORIA

The first step in the management of a case of esophoria is the correction of any existing error of refraction; and especially in a case of accommodative esophoria like the one before us, is it necessary to correct the hyyermetropia as fully as possible, and thus by lessening the innervation of the accommodation do we also check the tendency to excessive convergence.

Theoretically we would prescribe the lenses that represent the total amount of error we have been able to discover, but practically such lenses are too strong for comfort to start with, and, therefore, in this case, we will order + 1.50 D. spheres for constant wear.

Such lenses should be worn long enough for the eyes to become thoroughly accustomed to them in order to note what benefits are derived and what amelioration of the unpleasant symptoms, and as a rule, no attempt should be made to influence the muscles by means of prisms until all the improvement that could possibly follow from the spherical glasses has been exhausted.

When prisms are required for the correction of the esophoria, they are placed bases out. In such cases when the muscles are at rest, the visual lines assume the excessive convergence produced by the muscle imbalance and a homonymous diplopia is the result. Prisms bases out enable the eyes to maintain binocular single vision, thus avoiding the nervous strain which is otherwise unavoidable to maintain and not to exceed the proper amount of convergence.

If the esophoria does not exceed 2° or 3°, it will not call for correction by prisms. If of high degree, a tenotomy may become necessary, because on account of the distorting property of prisms, it is not pleasant to wear very strong glasses of this kind, 5° for each eye being the limit usually allowed.

The proportion of the esophoria that should be corrected by prisms varies in different cases, but, as a rule, it should not be more than one-half to two-thirds of that manifested by the test at 20 feet, which usually suffices to relieve the asthenopia, whereas a total correction would not be tolerated.

Prisms prescribed for the relief of esophoric asthenopia must ordinarily be worn constantly; occasionally in esophoria, which is not attended by spasmodic action of the internal recti, the tendency to excessive convergence disappears in near vision, and, under such conditions, it suffices to wear the prisms for distant use only.

In the slighter cases of esophoria much relief is afforded by the use of prismatic glasses; but in many cases the excess of convergence is so great that only a small proportion can be corrected within the limits allowed for such lenses. In some cases where relief is afforded at first, a greater amount of esophoria becomes manifest under the relaxing influence of the prisms, so that the strength of the latter must be increased until the limit is reached, and then other methods of treatment must be sought.

Cases of esophoria usually occur in connection with hypermetropia or some other form of ametropia, in which cases the prisms are combined with the refracting lenses, or rather the necessary lens curvature is ground upon the surfaces of the desired prismatic correction.

In this case we will combine prisms with the desired spheres, and especially as we do not feel justified in prescribing lenses strong enough for a full correction.

Our prescription will read as follows ·

$$\left.\begin{array}{l}\text{O. D.}\\ \text{O. S.}\end{array}\right\} + 1.50 \text{ D. S.} \supset \text{prism } 2° \text{ base out.}$$

This corrects a little less than half the esophoria, but on account of the unpleasantness (to the patient) of prisms, we prefer not to order them too strong at first.

Spasm of Accommodation

[CLINIC NO 30]

This patient, Miss Mary H., is twenty-five years of age. She tells us her glasses have been changed three times during the past year. The new glasses would seem satisfactory at first, but soon she was able to see better at a distance without them. Complains of pain in her eyes and headache whenever she attempts to use her eyes either with or without glasses. The pain in her head nearly sets her wild when she tries to concentrate her vision on a distant or a near object. Has been able to do but little reading, writing or sewing for the past year. Has been under treatment by specialists in gynecology and neurology, but they find but little departure from normal conditions. The suggestion has been made that she is hysterical. She sleeps well and has a good appetite, but occasionally suffers from nausea and vomiting if she uses her eyes for any great length of time. The patient is apparently in good physical condition, on account of which she receives but little sympathy from her family or friends, and she therefore almost seems provoked with her apparent good health as not being consistent with her suffering.

You will remember the first step in the actual examination of a case is to ascertain the acuteness of vision, which in this case we find to be $\frac{20}{70}$, being the same for each eye.

We next measure the amplitude of accommodation, which we find to be 10 D., depending upon a near point of four inches, as measured from the eye to the reading card, which is approximated as close as it is possible to decipher the small type at the top of the card.

We turn to the trial case test and find convex lenses are rejected as blurring the distant vision. We commence with $+ .25$ D., which is refused; in spite of this we try stronger lenses, but each increase of strength blurs vision still more. We try each eye separately and find the result the same.

There seems to be nothing else but to try concave lenses, which we will proceed to do, but only for the purpose of pointing a moral. These lenses are quickly accepted, and we soon find that $- 1$ D. raises the vision to normal.

We now test the muscle balance by means of the Maddox rod, according to our usual custom. The rod being placed over the left eye, the patient tells us the streak is about six inches to the left; after a few trials we find that a prism of 4° base out is required to bring the streak back to the flame, which then represents the amount of esophoria. We repeat the test with the — 1 D. lenses in the trial frame: now the patient says the red streak is farther to the left and a 5° prism base out is required to bring it back. The correction of the apparent defect by the concave lenses has therefore added to the muscle disturbance and increased the esophoria.

VERIFYING THE DIAGNOSIS

Presumably then we have a case before us of simple myopia, but we must hesitate and give the case further consideration before we can accept this diagnosis as correct.

In the first place the symptoms of which the patient complains do not indicate myopia. They are symptoms of asthenopia, which does not commonly occur in connection with myopia.

In the second place, the muscle imbalance is one of esophoria, which is usually associated with a hypermetropic condition of refraction, and besides this imbalance is aggravated by the lenses which the patient accepts.

In the third place, hypermetropia is the predominant error of refraction, and for this reason, as well as to give the patient the benefit of any doubt that may exist, we should always approach a case suspecting hypermetropia and make our examinations such as to discover hypermetropia if possible, and not to abandon this supposition until it is positively proven that hypermetropia is not present.

In the fourth place, when the refraction seems to be myopic, we should always suspect the possibility of false or apparent myopia, especially in a young person like this, and where the symptoms are such as to indicate some nervous disturbance.

In the fifth place, the possibility of spasm of the accommodation should always be kept in mind, in persons under middle age, and especially where the glasses have been changed frequently as in this case, with but little satisfaction after the several changes.

In view of these considerations we must look upon the diagnosis of myopia with great suspicion, and make our tests rather with a view of uncovering hypermetropia. If the latter is not present our examination can do the patient no harm; but if present in a latent form we can do the patient much good by detecting its presence, in fact, this is the only way we can do our whole duty conscientiously for the patient's welfare and our own reputation.

Inasmuch as the two eyes are so nearly alike, we will test them together in the act of binocular vision, as in this way it is possible to secure better results. We will place in the trial frame before the patient's eyes a pair of $+ 4$ D. lenses. Before she has an opportunity to complain that she cannot see through them, we will tell her that we expect the glasses to blur her vision very much, but that we use them to accomplish a certain purpose, and that therefore we will ask her to submit to their inconvenience for a little while.

You gentlemen will of course understand that we are using these somewhat strong convex lenses to repress the accommodation, which is probably unduly active, and in this way we hope to discover the real condition of the eye which may be very different from the manifest condition, as we have found it. In order to secure the relaxing effect of these convex lenses on the accommodation, we will allow the lady to wear these lenses for a little while, and in the meantime I will call your attention to a few facts about

SPASM OF ACCOMMODATION

The constant strain on the ciliary muscle that is necessary to afford clear vision in hypermetropic conditions of refraction, gives rise to spasm of the accommodation, which is simply a continued and persistent contraction of the ciliary muscle.

Such spasm may occur in any condition of refraction, even in emmetropia, which is then transformed into an apparent myopia. When it occurs in hypermetropia, the defect is concealed and the eye made apparently emmetropic, or the contraction of the muscle may go too far and the eye made apparently myopic, as we suspect in the case before us.

A simple hypermetropic astigmatism by this means is transformed into a simple myopic astigmatism, the defective meridian being transposed to right angles. A myopic eye is made still more myopic.

Spasm of the accommodation has been the bug-a-boo of the optometrist, because it is so difficult to detect and conquer it without a mydriatic. In the majority of cases, if the causes can be discovered and remedied, the spasm will gradually disappear, and the expert optometrist will be able to accomplish this without resort to a mydriatic.

Spasm of accommodation is more apt to occur in persons of a nervous temperament, and strange to say it does not depend upon the vigor of the accommodation; that is to say, persons with a relatively feeble accommodation may suffer from cramp of the ciliary muscle more than persons of strong muscular development. It is usually found in connection with a weakened accommodation, and instead of being an evidence of strength, must be regarded as an indication of nervous debility.

The causes of accommodative spasm are many and various. The direct cause is the necessary effort of the ciliary muscle to overcome a hypermetropic condition of refraction, and I may also mention a trial of sight by concave lenses as a contributing cause.

The indirect causes are a neurotic condition of the patient, a systematic tendency to spasmodic affections, a hyper-sensitive retina, overwork or abuse of the eyes and muscular insufficiency.

SYMPTOMS OF SPASM OF THE ACCOMMODATION

The *symptoms* of accommodative spasm are photophobia (dread of light), lachrymation (excessive watering of eyes), pains or discomfort of some kind, contracted pupils, hyperæmia of the conjunctiva, impairment of distant vision and simulated myopia.

A certain amount of spasm of accommodation is really physiological and is due to the normal tone of the ciliary muscle. The condition of rest of the ciliary muscle is not one of complete relaxation, as in a state of paralysis. The tone of the muscle of accommodation keeps it in a condition of mild contraction.

This causes a slight increase in the refraction of the eye, amounting to 1 D. or more in childhood, and diminishing to .50 D. or .25 D. in middle age. Hence even in eyes apparently emmetropic when atropine is used, the refraction of the eye is slightly less than when tested in its normal condition.

It is only when the ciliary muscle (which may be said to be extremely excitable) exceeds this physiological limit and keeps up its excessive action constantly, that we have the condition which we are considering and which we know as spasm of accommodation.

In testing the vision of a patient with accommodative spasm, you will be likely to find great variability in the acuteness of vision.

At first he may say he can only see the larger letters at the top of the card, then the smaller letters begin to come into view, and he reads down almost to the 20 line, which is in turn again blotted out and only the larger letters remain legible. The varying acuteness of vision is dependent upon the contraction and relaxation of the ciliary muscle.

Another evidence of spasm of accommodation is a variableness in the appearance of the radiating lines on the clock-dial card, at one time patient saying one set of lines appear clear and distinct, and the next moment they become dim and another set of lines appear plainer.

In simple hypermetropic astigmatism, there is apt to be spasm of accommodation, which transposes the case into myopic astigmatism, with the defective meridian changed to right angles.

In hypermetropic astigmatism with the rule, the horizontal meridian is hypermetropic and the vertical meridian emmetropic. The ciliary muscle instinctively comes into action, neutralizing the hypermetropia in the horizontal meridian and making it apparently emmetropic, and at the same time increasing the refraction of the vertical meridian to an equal extent and making it apparently myopic. In this way the defective meridian is changed from horizontal to vertical, and the location of the indistinct lines would vary with the contraction and relaxation of the accommodation.

Another evidence of accommodative spasm is the variation in the glasses accepted by the patient at different examinations, or perhaps even during the same test.

Corroborative evidence of spasm of the accommodation is furnished when the retinoscopic findings vary greatly from the test with the trial case. In the first case the darkened room and the request to patient to look off into distance without fixing the sight on any one object in particular, presents conditions that favor relaxation of any spasm of the ciliary muscle that may be present.

DIAGNOSIS OF SPASM

The diagnosis of spasm of accommodation can be made in accordance with the characteristics I have outlined to you, and corroborated by a determination of the real refractive condition by persistent and patient fogging, or in extreme cases by the physician's aid in the use of a cycloplegic, which in obstinate cases must sometimes be used for several days or a week. I might say in passing that if atropine is used the proper strength is four grains to the ounce, or approximately a 1 per cent. solution. It may be dropped into the eyes two or three times a day, until the physiological effect of the belladonna becomes apparent in the flushed face and the parched throat.

A diagnostic point of some value is a discrepancy between the position of the far point and the degree of apparent myopia as shown by the test lenses.

For instance the far point may be thirteen inches, which would indicate a myopia of 3 D. The acuteness of vision is $\frac{20}{50}$, which is easily raised to normal by — 1 D. With such a marked difference, you would at once suspect accommodative spasm.

TREATMENT OF SPASM

In the treatment of spasm of the accommodation, the first step is to discover and remove the cause, but I must acknowledge that the diagnosis of the cause of the spasmodic action is not always easily made. I have already referred to the causes of spasm, and can only repeat that the effort to overcome a refractive error is the most common cause; in addition to this it may be dependent upon an insufficiency of convergence, which is to be explained as follows: the excessive effort required to maintain the necessary convergence excites a corresponding effort of the accommodation, on account of the close relation that naturally exists between the accommodation and convergence. Under such circumstances a pair of prisms bases in would assist in relaxing the ciliary spasm by removing the cause, which is the desire for excessive effort of the internal recti muscles.

If the eyes have been overworked, they should be rested if possible for a few days before the examination is made, during which time of rest light smoke glasses may be worn and especially if the retina is hyper-sensitive.

If a conjunctivitis is present or there is a noticeable nervous element in the case, the services of a physician should be availed of in the treatment of such conditions, in order to place the eyes in a condition more favorable for examination.

The direct repression of the excessive accommodation is accomplished by means of convex lenses, a moderately strong pair for reading and a weaker pair for distance. Of course, wearing convex lenses for this purpose makes vision indistinct, and the patient is apt to rebel. The *rationale* of the treatment must be explained to him, and in cases of intelligent persons their co-operation can be secured. Otherwise the distance glasses will probably have to be dispensed with and reliance placed on the reading glasses, to which few persons will object.

The rule as set forth in all text books to give the strongest convex lens with which patient can read the No. 20 line, or the weakest concave lens that affords the same vision, is a time-honored one, and shows recognition of the fact that it is always desirable to relieve the ciliary muscle or impose as slight a tax upon it as possible.

But fogging for the uncovering of latent hypermetropia is of later origin, whether it be used temporarily during the examination, or more permanently by prescribing such glasses for constant wear, which not only tend to check ciliary innervation, but also probably by dulling the vision relieve the sight centers of the tax which sharp vision entails. In some parts of the world where the atmosphere is humid and foggy the greater part of the time, the statement is made that nervous ailments are uncommon, because in a hazy atmosphere the visual centers are not taxed to the same extent as in a clear atmosphere. Fogging glasses sometimes produce a state of quiet and rest even to the extent of sleep.

We will now return to our patient who has been very patiently waiting, and she tells us the glasses are not so uncomfortable as at first, and that she can see a little better. On directing her attention to the test card, she is barely able to name the largest letter at the top of the card, showing a vision of $\frac{20}{200}$.

We cautiously and slowly reduce the convex lenses by concaves placed in front, commencing with $-.50$ D., which affords an improvement, and increasing .25 D. at a time until -2 D. is reached, when the No. 20 line is perfectly clear and legible. We therefore in this case have uncovered 2 D. of hypermetropia.

We will order $+1.50$ D. for constant wear, which from our experience in similar cases, we feel confident will afford the greatest relief and comfort, if the lady will overlook the slight annoyances she may feel during the first week, and persevere in their use.

This case then is one of apparent or accommodative myopia; it is as you see not myopia at all, but it has been made to simulate myopia on account of the spasm of accommodation which has been present to overcome the hypermetropia, which is the real refractive condition of the eye. The innervation of the ciliary muscles becomes excessive, and passes beyond that point where it would suffice to neutralize the diminished refraction, and in going farther produces the simulated myopia.

I hope this case has made such a deep impression upon your minds that you will carry it away with you, and that the lessons you have learned from it will be of value to you in your daily practice.

How easily concave glasses could have been given to this lady, and how injuriously they would have affected her. Always be on your guard in your patients for spasm of accommodation, and always be slow in prescribing concave lenses. Carry this piece of advice with you: *Suspect every case to be hypermetropic, until positively proven otherwise.*

Exophoria

[CLINIC No. 31]

This little patient is Master A. E. K., aged nine. His mother tells us that he has complained of headaches ever since he was three years of age. She has just consulted an oculist about him, who recommends an operation as the only method of treatment that will afford relief. She objects to this, however, and brings the child to us to see if he cannot be fitted with glasses that will be of benefit to him.

We find the acuteness of vision of each eye to be $\frac{20}{30}$ partly.

We turn to the ophthalmometer, which shows no overlapping in any meridian, proving all the meridians to be of equal curvature. Now, inasmuch as the normal cornea shows a slight excess in the vertical meridian (about .50 D.), which is lacking in this boy, we must class this case as one of astigmatism against the rule, that is, about .50 D. deficiency in the vertical meridian.

Some students have difficulty in understanding how astigmatism is indicated when the mires maintain the same relation as the ophthalmometer is rotated, showing neither overlapping nor separation in any meridian. To their minds this rather indicates absence of astigmatism. But it must be remembered, and this is an important point, that normality of the corneal curvatures is not equality in all meridians, but an excess of .50 D. in the vertical meridian. If now the vertical meridian is on an equality with the horizontal, when it should exceed it, we must conclude that its refraction is deficient, which means astigmatism against the rule.

If the ophthalmometer shows the vertical meridian to exceed the horizontal by .50 D., we say no astigmatism. If it exceeds by a greater amount than .50 D., we say astigmatism with the rule.

We next make a retinoscopic examination, which shows the horizontal meridian to be emmetropic, and the vertical meridian to be hypermetropic to the extent of .50 D.

The retinoscope and the ophthalmometer agree, and we have a plain case as far as the condition of the refraction goes. We now turn to the trial case to see if the subjective examination will verify the objective examinations. We try $+$ .50 D. spheres, which the boy does not like very much, but when we replace them with + .50 D. cyl. axis 180°, he says at once, the last are the best. We try stronger cylinders, and we try the addition of spheres, but both are rejected. We rotate the cylinder, first one way and then the other, with the effect of making vision worse, showing 180° to be the proper position for axis of cylinder. We have thus verified the correction of the refractive error.

In the regular course of our examination we come now to look into the muscle balance. We make use of the Maddox rod as is our usual custom, placing it over the left eye. We explain to our little patient that he will see the natural light, and in addition he will see a red streak of light. In order to make this red streak more noticeable, we rotate the Maddox rod, and by this means he is the better able to locate the streak. He tells us the red streak is away over to the right. This means crossed diplopia due to exophoria. This is to be corrected by prisms, bases in, and we try prisms, gradually increasing their strength, until we find 12° are required to bring the streak and the light together, and this then represents the amount of the exophoria. The muscle imbalance is the essential feature of this case, for which the oculist previously consulted had advised an operation. It will be profitable for us to give a little time to-day to the consideration of this condition of exophoria.

Deficiency of convergence may be latent, as in the present case, when we call it exophoria, or it may be manifest when it is known as divergent strabismus. We use the words "deficiency of convergence" advisedly, as implying that exophoria in most cases is to be regarded as a lack of convergence rather than an excess of divergence. It is a lessened innervation of the function of esophoria, where there is convergence, thus contrasting with the conditions present in an excessive innervation of the convergence.

SYMPTOMS OF EXOPHORIA

Exophoria is the condition that was formerly known as "muscular insufficiency," and later as the muscular form of asthenopia, which occurs especially when the eyes are tired from prolonged near work and particularly by artificial light.

Headache is a marked symptom, and you will recall that this boy has suffered from headache ever since he was three years of age. This headache is naturally aggravated by close use of the eyes, and is sometimes accompanied by such reflex symptoms as dizziness, nausea, vomiting and fainting. Or the eyes may feel weak and tired, the words "seem to jump," or the letters "run together," sight grow dim or objects appear double for a moment, without the customary headache.

Sometimes near work must be abandoned on account of the disturbance produced by over-taxation of the internal recti muscles in their effort to maintain convergence, or by an impossibility to maintain convergence resulting in confused vision or crossed diplopia.

CAUSES OF EXOPHORIA

As we have found esophoria in the majority of cases to be dependent upon hypermetropia, so on the other hand is exophoria associated with a myopic condition of refraction, and for the same reasons. In the first case the accommodative effort required to overcome the hypermetropia, stimulates the convergence to extreme effort. In the second case, the myopic eye being adapted for near vision, requires but little accommodative impulse, and hence the convergence center lacks the customary stimulus, thus giving rise to an insufficiency of convergence or exophoria.

But in this boy's case we have found the refraction to be slightly hypermetropic (and astigmatic), and hence we must look for some other cause. In the further investigation of the muscles, we will measure the power of the convergence and of the divergence, which is accomplished by the strongest prisms bases out and in respectively that can be overcome.

We place a 2° prism base in before the right eye, while we ask the boy to look at the light and tell us if he sees one or two. He replies that he sees only one. We place another 2° prism base in before the left eye, and he tells us that he still sees one light. We increase first one prism and then the other to 4°, and still the light remains single. We increase one prism to 6°, and now he says he sees two lights; we diminish to 5° and still he sees two; we return to 4°, with which there is a single light. We have now two 4° prisms before the eyes, which represents a divergence power of 8°, which is the full normal amount, but not any excess.

We now place a prism of 2° base out before the right eye; this at first causes two lights to be seen, but they quickly fuse into one. We place a 2° prism base out before the left eye; two lights are again seen, which with some effort are fused into one. We increase the right prism to 4°, with the result of producing a diplopia which the boy is unable to overcome. We return to 2°, with which the light remains single. We have now 2° prisms before each eye, which represents a convergence power of only 4°, which is very much below the normal standard.

In this case then the exophoria is not due really to an excess of divergence, but to a very great diminution of convergence. There is only a seeming excess of divergence because the convergence is so much below the standard. The external recti muscles are relatively strong, but not absolutely so.

This case is then literally one of "insufficiency of the internal recti," and may be due to actual weakness of these muscles, or inaction of these muscles because of imperfect innervation. In a case like this we think an operation is contra-indicated as likely to do more harm than good.

When you have charge of a case of this kind, it is very difficult to determine the proper proportion that should be observed between rest and exercise, in giving advice as to the use of the eyes. It only seems reasonable that rest of a fatigued organ would afford relief. Inasmuch as it is impossible to close the eyes to obtain entire rest, we do the next best thing by a correction of the refraction, which affords relief to the ciliary muscle, and prisms to rest the extraocular muscles. Then we advise our patient that near use of the eyes must be discontinued just as soon as they begin to feel tired.

DIAGNOSIS OF EXOPHORIA

The simplest method is to produce an artificial vertical diplopia by means of a prism placed base up or down. We will take an 8° prism and place it before this boy's right eye base up, while we ask him to look at the light across the room. He sees two lights, one below the other. We know the lower light belongs to the right eye, because a prism always displaces an object towards its apex. But in order that the patient may distinguish between the two lights the more readily, we will make use of a red glass over the right eye.

The image formed in the left eye is the natural light, and that formed in the right eye is red. We ask the patient if he sees one natural light and one red light, and which light is above and which below. He replies yes, and that the lower light is the red one.

This we know to be correct, but in our further questions we must depend upon the patient's answers.

"Is this red light directly below the white light, or is it to the right or left?" is the next question.

"It is way off to the left."

"About how far to the left?"

Our little patient seems unable to answer this question, but he holds his hands up indicating a distance between them of about two feet.

This then is a case of crossed diplopia, which we know must be due to exophoria, and we proceed to measure the amount of same by determining the degree of prism that is necessary to bring the red light back in a line with the white light. This we finally find to be 12°. The base of the correcting prism is *in*.

As I stated a moment ago this is the simplest test, but it is subject to one disadvantage, and that is a displacement from the yellow spot of the image formed in the right eye by the strong vertical prism placed in front of it.

Usually I prefer the Maddox rod test. You are all familiar with this little instrument. The one I hold in my hand is a multiple rod, being a series of parallel cylinders mounted in an opaque diaphragm of suitable size to be placed in a trial frame. These cylinders produce a marked elongation of images in a direction at right angles to their axes. A small flame seen through this multiple rod appears as a long red streak of light.

In testing the muscles to detect the presence of exophoria, the rod is placed in the trial frame in a horizontal position, in order to produce a vertical streak of light. The patient receives the image of the streak in one eye, and of the uncovered light in the other, and on account of the dissimilarity in the shape, size and appearance of the two retinal images, it is impossible, in fact, nature makes but little effort to fuse them, and in abandoning the attempt to produce fusion, the eyes are free to assume their position of equilibrium, and in this way any tendency to deviation which has been latent now becomes manifest.

We place this Maddox rod over the boy's right eye, and in answer to our question he says the red streak is considerably to the left of the light. A prism base in brings it closer, and after a few trials we find that a 13° prism causes the streak to pass vertically through the flame.

This result does not exactly correspond with the former test, but we must expect some variation from the several tests, or even from the same test repeated at different times.

Now, it is a well-known fact that when exophoria exists at twenty feet, it is of higher degree at the reading distance. And even in some cases which show slight esophoria at distance, there will be exophoria at the reading point. This is due to the fact that exophoria, being caused by an insufficiency of the internal recti muscles, will show of higher degree at the reading distance where the tax on the insufficient muscles is greatest. This is called *exophoria in accommodation*, which term is used to denote the insufficiency found at the reading point, in contrast with the simple word exophoria, which indicates the imbalance at a distance.

We will bring this boy's chair close up to the light, and repeat the test in order to determine the amount of exophoria in aecommodation. We place the Maddox rod in the same position, and the boy sees the red streak to the left, and we find a 20° prism base in is required to bring the streak back to the light. This serves to emphasize the weakness of the internal recti muscles that is present in this case.

TREATMENT OF EXOPHORIA

1. *Correction of Refraction.* As ametropia is regarded as being the most common cause of muscular insufficiency, it is obvious that the first step in the treatment of such insufficiency should be the correction of any existing error of refraction. Sometimes such lenses will suffice to restore a proper muscle balance; but at any rate they remove a disturbing factor which would otherwise stand in the way of a cure.

In this boy's case we will order the weak convex cylinders which we found in the early part of our examination. If his eyes were myopic, as is usually the case in exophoria, for the reasons already explained, we would order the correcting lenses alone in the expectation that they would beneficially affect the convergence; but we can hardly look for any such improvement from these weak convex cylinders, and hence we must use prisms in the treatment of this case.

2. *Prismatic lenses.* Prisms are often necessary for the relief of exophoric asthenopia, within the limitations to which this form of lenses is restricted. The prism is placed in such position as to

afford assistance to the weakened and strained convergence, which in these cases would be *base in.*

Exophoria is almost invariably greater at the reading distance than at twenty feet, as we have seen in this case, and for this reason one strength of prisms will not always suffice for constant wear for all purposes. Sometimes where the asthenopia is noticed only after prolonged near work, it may suffice to wear the prisms only in close use of the eyes; but if the exophoria is of high degree, relief can only be obtained by the constant wearing of prisms, in which cases stronger prisms may be necessary for prolonged close use. The strength of the prism to be prescribed is always an open question, that must be decided by a consideration of each case individually. It is never proper to give prisms strong enough to correct the full amount of the exophoria, but as a general rule I would say about one-half the correction should be given, which in exceptional cases may be increased to two-thirds.

Some authorities claim that prisms for constant wear do more harm than good, by increasing the difficulty and requiring stronger and stronger prisms. I cannot agree with these views, because by such use of prisms we relieve the great strain upon the convergence, which function is then called upon to perform only so much work as it can comfortably do, and in this way is placed in a position to regain its lost strength or even develop more power.

I will write on the blackboard the prescription I will order for this boy: O. U. + .50 D. cyl. axis 180° ◯ prism 3° base in. This represents a total prismatic value of 6°, which is just one-half the amount of the exophoria manifest at twenty feet. In a boy as young as this, we can scarcely order stronger prisms for reading on account of the inconvenience of two pairs of glasses.

Prism Exercises. I do not feel that I can close the clinic without giving you a few directions on prism exercises to develop the strength of the convergence. The prisms are placed bases out, and the strongest pair is found with which singleness of vision with the distant light can be maintained. These are set in a frame and worn for five or ten minutes, being lifted about every thirty seconds. This exercise may be repeated daily, the effort being made to increase the prisms each day. Very weak prisms will be necessary to start with, and the exercises should be continued until the power of adduction has reached 20° at least; in some cases it may not be difficult to reach 40 or 50°.

These exercises you will have to conduct yourself in your office, but in addition you may give your patient for home use weak prisms set with bases out. They should not be too strong and they may be worn for an hour or two (continuously or at intervals, as the eyes best stand them) every day as the person is doing his daily work.

These exercises may be varied as follows: a pair of weak prisms such as can be easily evercome is placed on the patient's eyes bases out, while he is standing two feet from the light. He is requested to recede from the light slowly while keeping his eyes fixed upon it. If diplopia shows itself at any distance, he is to return to his original position and recede again until he can get to twenty feet without diplopia. The exercise may then be repeated with slightly stronger prisms, and then finally he remains at twenty feet and raises and lowers the prisms while gazing at the light, as previously directed.

In addition to these prism exercises, the patient may carry out systematic exercises of the muscles (called ocular gymnastics), somewhat as follows:

Tell the patient to look intently at the point of a pencil held at arm's length, and then slowly bring the same closer to the eyes. If the pencil becomes double, it should be pushed away and the exercise repeated from the original distance. This procedure is one that has the merit of convenience, as it can be practiced at any time or place and requires no apparatus.

Transposition of Lenses as Illustrated by a Case of Mixed Astigmatism

[CLINIC NO. 32]

E. M. G., aged forty-eight years, a professional man. Complains of some difficulty in both distant and near vision, also more or less discomfort in head and eyes. Has worn glasses for the last twenty-five years, but as they have not been changed for four years, he feels they are no longer suitable.

We find his acuteness of vision is $\frac{20}{80}$ partly, and the same for each eye.

We turn to our trusted ophthalmometer, the readings of which show an overlapping of 2.50 D. in the horizontal meridians. For reasons which have already been explained to you at former clinics, this case proves to be one of astigmatism against the rule, and making allowance for the proper additions, the amount of astigmatism is 3 D., the excess of curvature being in the horizontal meridian.

We take up the trial case examination next, and commence with a $+$ 1 D. cylinder, which according to the indications of the ophthalmometer, we place with axis at 180°. This causes an improvement in vision, and the next line now becomes readable. We try to increase the strength of the cylinder, but without success.

As we have apparently corrected one meridian, and as the ophthalmometer shows a still greater amount of astigmatism, we suspect it must be in the other meridian, and leaving the convex cylinder in place we add to it a concave cylinder with axis at right angles. We commence with a $-$ .25 D. cyl. axis 90°, which affords a noticeable improvement in vision, and increase 25 D. at a time, each change of lenses producing still clearer vision. When we reach a $-$ 1 D. cylinder, the patient is able to quickly name all the letters on the No. 20 line. We test the other eye and find exactly the same correction.

This then is a case of mixed astigmatism against the rule, the degree of defect being represented by the following cross cylinder :

$+$ 1 D. cyl. axis 180° $\bigcirc$ $-$ 1 D. cyl. axis 90°,

This gentleman complains that he has considerable difficulty in the close use of his eyes, and that reading has lost all attraction for him ; and no wonder, for he has already reached the presbyopic age, and the glasses that correct his error of refraction no longer suffice to afford the necessary help in close use of the eyes.

We place these cross cylinders in the trial frame, and hand the reading test card. He holds it at arms' length, and even then he can scarcely make out the smaller size print. At this age (forty-eight years) we usually expect to find from 1 D. to 1.50 D. of presbyopia, and we will therefore add a $+$ 1 D. lens to the cross-cylinders. This produces a marked improvement in reading vision, and makes the type sharp and clear at the customary distance.

We will of course make a test of the muscle balance, and finish up with an examination by the ophthalmoscope and retinoscope, but we will not dwell on these matters to-day, as our special interest in this case lies in the transposition of lenses involved in these prescriptions.

TRANSPOSITION OF LENSES

Transposition means a change in the curvatures of a lens without affecting its refractive value, and is a matter of mathematical calculation as well as algebraic addition. Preference is usually given to the simpler and less expensive form of lens, and therefore when an optometric examination results in a cross-cylinder as in this case, it is customary to transpose the formula to a sphero-cylinder before sending the prescription to the manufacturing optician.

This formula is one in which the two cylinders are of dissimilar signs and with their axes at right angles to each other. In such cases the rule is as follows : <u>Take either one of the cylinders for the sphere of the new combination, retaining its sign ; make the cylinder of the new combination from the sum of the two former cylinders, using the sign and axis of that cylinder which was not used for the sphere.</u>

Let me write the formula on the blackboard and then proceed to transpose it before your eyes so that you can follow me ·

$+$ 1 D. cyl. axis 180° ⌒ $-$ 1 D. cyl. axis 90°.

We will take the first-named cylinder for our sphere, retaining its sign ; this gives us $+$ 1 D. S.

Then we will take the sum of the two cylinders for our new cylinder. Now you will please note the words I use, viz., the

Transposition of Lenses in a Case of Mixed Astigmatism

sum of the two cylinders, which in this case would be : 1 and 1 = 2. I do not say the addition of the two cylinders, because the algebraic addition of + 1 and — 1 would equal nothing.

But instead it is really algebraic subtraction, which is the process of finding the difference between two numbers, or to express it in other words, the number of units which lie between the two numbers.

Now in this formula under consideration, the plus number represents a unit on the positive side of zero, and the minus a unit on the negative side of zero, and the difference between them is two units.

Where the numbers to be subtracted have dissimilar signs, that is, where one is plus and the other minus, then the subtraction really means addition in accordance with the following rules :

When a positive number is to be subtracted from a negative number, we change the sign to minus and proceed as in addition, and the result is a minus number.

When a minus number is to be subtracted from a positive, we change the sign to plus and proceed as in addition, the result being a positive number.

For those members of the class who have never studied algebra, these few remarks on algebraic subtraction will prove of value, not only in this case, but many times when a transposition becomes necessary or desirable.

Now to return to the formula under consideration, we already have the sphere of the new sphero-cylindrical combination, and we obtain the cylinder by subtracting the first cylinder from the second, according to the rule I have already given you, and the result which I will mark on the blackboard will be as follows :

Subtract + 1 from — 1 = — 2.

I will repeat the rule again so as to fix it in your memory · change the sign of the subtrahend (the number to be subtracted) and proceed as in addition.

We have now for the new formula + 1 D. S. ◯ — 2 D. cyl., and of course the new minus cylinder retains the axis of the old minus cylinder, which in this case is 90°.

There is a second transposition we can make according to the same rule, as follows : this time we will take the minus cylinder for our sphere (— 1 D. S.), and then we subtract this — 1 from + 1, which equals + 2, retaining of course the original axis of the convex cylinder, the result being as follows : — 1 D. S. ⌒ + 2 D. cyl. axis 180°.

Now if you will look on the blackboard you will see the original cross-cylinder and the two sphero-cylindrical transpositions made from it.

+ 1 D. cyl. axis 180° ⌒ — 1 D. cyl. axis 90°
+ 1 D. S. ⌒ — 2 D. cyl. axis 90°
— 1 D. S. ⌒ + 2 D. cyl. axis 180°

These three formulæ are all inter-transposable, and they all have the same optical value and the same effect on the rays of light passing through them. Such being the case the question may occur to you, which one is preferable?

It is seldom that a cross-cylinder is ordered; it is customary to transpose it into a sphero-cylinder, and of the two sphero-cylinders into which this cross-cylinder is transposable, the first is the one to be preferred, and why?

Now you understand that the axis of a cylinder is plane glass, the refractive power being in the meridian at right angles, where the effect is the same as a sphere of like power.

Now you also understand that when we look through a sphere at any place except its optical center, a prismatic effect is produced.

And you will further remember from your studies of the extraocular muscular system that the vertical muscles are more easily disturbed and thrown out of balance than the stronger horizontal muscles.

Now bearing these three facts in mind, let us consider the advantages and disadvantages of these two sphero-cylinders.

In the first one the axis is at 90°, and therefore there is nothing to disturb the vertical muscles, while the slight prismatic effect produced as the eyes are turned from side to side is easily taken care of by the external and internal recti.

In the second sphero-cylinder the axis is at 180°, which throws the refractive power of the cylinder in the vertical meridian, producing a prismatic effect as the eyes are turned down (and are we not constantly turning our eyes downward more or less), and in this way disturbing the balance of the vertical muscles, which are relatively weak and but little able to overcome the prismatic effect without showing signs of asthenopia. Therefore you can under-

Transposition of Lenses in a Case of Mixed Astigmatism 211

stand the advantages of the first sphero-cylinder, which imposes no strain on the vertical muscles.

One of the gentlemen tells me he cannot clearly comprehend the reason of the transpositions I have given you or why the given results are obtained, and as there may be others who are a little at sea on the matter, I will give you a further explanation, by means of diagrams on the blackboard.

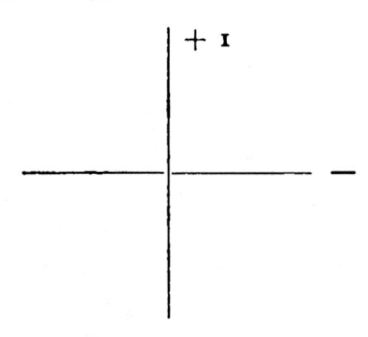

$$\begin{array}{c} + 1 \\ \underline{\quad\quad\quad\quad} \; - \\ \end{array}$$

In the cross cylinder there was plus one power in the vertical meridian and minus one power in the horizontal, always remembering that the power lies at right angles to the axis, and any transpositions that are made must strictly retain the same powers in each meridian.

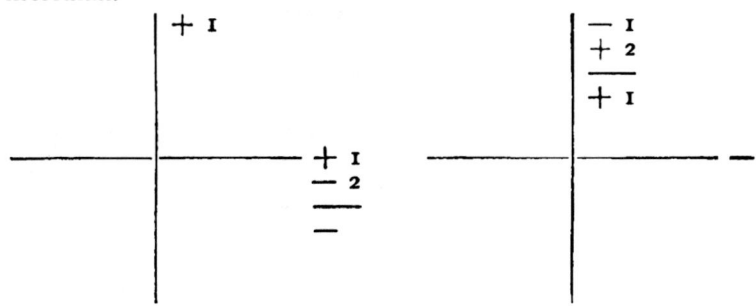

$$\begin{array}{cc} +1 & \begin{array}{c}-1 \\ +2 \\ +1\end{array} \\ \underline{\quad\quad} \; \begin{array}{c}+1 \\ -2 \\ -\end{array} & \underline{\quad\quad} \; - \end{array}$$

In the first sphero-cylinder there is a $+ 1$ D. sphere, which gives its power in both meridians, and which I will mark on the diagram. Then we have a $- 2$ D. cylinder with its axis at 90°, and its refractive power lying at 180°. Now then a $+ 1$ placed against a $- 2$, according to algebraic addition yields a $- 1$. A comparison with the first diagram shows we have the same values in each meridian in this sphero-cylinder.

In the second sphero-cylinder there is a $- 1$ D. sphere which gives a $- 1$ value in both meridians as marked on the diagram. Then we have a $+ 2$ D. cylinder with axis at 180° and refractive power at 90°. The addition of a — 1 and a + 2 algebraically equals + 1. A comparison of this diagram with the other two diagrams shows the same power in both meridians in all three cases.

For the reasons mentioned we will order the first-named sphero-cylinder for constant wear, and for reading + 1 D. sphere to be added. I will write the formula on the blackboard so that you can all follow me.

$$\begin{array}{l} + 1 \text{ D. S. } \supset - 2 \text{ D. cyl. axis } 90°. \\ \underline{+ 1 \text{ D. S. added}} \\ + 2 \text{ D. S. } \supset - 2 \text{ D. cyl. axis } 90°. \end{array}$$

We tell the patient he can have his glasses made in bifocal form, otherwise he must have two pairs. He says that he does not like bifocals and that he prefers to have two pairs of glasses. His reading correction then will be as I have marked on the board, a + 2 D. sphere combined with a — 2 D. cylinder.

This presents another opportunity for transposition in order to reduce this formula to its simplest form. The — 2 D. cylinder neutralizes the + 2 D. sphere in one meridian, and leaves + 2 D. value in the meridian at right angles. In other words we may say that when the sphere and cylinder are of like amount with dissimilar signs, they equal a plane cylinder.

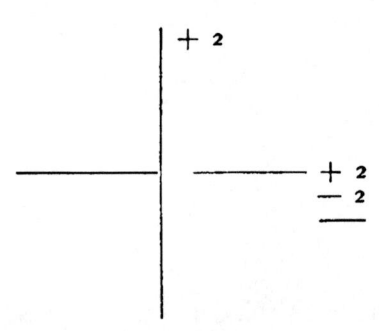

$$\begin{array}{c} \mid + 2 \\ \mid \\ \underline{\qquad\qquad\mid\qquad\underline{\quad + 2}} \\ \mid \qquad\qquad\quad - 2 \\ \mid \\ \mid \end{array}$$

Perhaps I can make my meaning clearer by a diagram on the blackboard. The + 2 D. sphere gives a + 2 D. power, in both horizontal and vertical meridians. The — 2 D. cyl. with axis at 90° gives a — 2 D. power in the horizontal meridian and plano in the vertical meridian. As you see the addition of + 2 and — 2 in the horizontal meridian yields nothing, or in other words, the one neutralizes the other in this meridian, while the + 2 D. power is left unaffected in the vertical meridian. Therefore the proper transposition for reading would be + 2 D. cyl. axis 180°.

Transposition of Lenses in a Case of Mixed Astigmatism 213

For sake of verification we may vary the presentation of the problem as follows :

$$\frac{\begin{array}{ll}+\ 2\ \text{D. vertical} & +\ 2\ \text{D. horizontal} \\ 0 & -\ 2\ \text{D.} \quad\text{``}\end{array}}{+\ 2\ \text{D. vertical} \qquad 0}$$

the result being exactly the same.

While I am on this subject of transposition, let me give you the rule for the transposition of a sphero-cylinder, according to which any sphero-cylinder can be transposed.

The *sphere* is obtained by the algebraic addition of the sphere and cylinder.

The *cylinder* is retained as in the orignal except that its sign and axis is changed.

In order to illustrate this rule I will write the formula of a sphero-cylinder on the blackboard, and then proceed to transpose according to this rule.

$$\frac{\begin{array}{ll}+\ 1\ \text{D. S.}\ \supset\ +\ .50\ \text{D. cyl. axis } 90° \\ +\ .50\ \text{D.} \qquad \text{change sign and axis}\end{array}}{+\ 1.50\ \text{D. S.}\ \supset\ -\ .50\ \text{D. cyl. axis. } 180°.}$$

Here we have a $+\ 1$ D. sphere and a $+\ .50$ D. cylinder with axis at 90°. For the new sphere we add the sphere and cylinder together, and as both have plus signs the result is $+\ 1.50$ D. For the cylinder we change the sign from $+$ to $-$, and the axis from 90° to 180°, the resultant sphero-cylinder being shown on the blackboard.

Now we will take this same sphero-cylinder and change it back again as follows

$$\frac{\begin{array}{ll}+\ 1.50\ \text{D. S.}\ \supset\ -\ .50\ \text{D. cyl. axis } 180° \\ -\ .50\ \text{D.} \qquad \text{change sign and axis}\end{array}}{+\ 1\ \text{D. S.}\ \supset\ +\ .50\ \text{D. cyl. axis } 90°}$$

In this case the addition of the $+\ 1.50$ D. and the $-\ 50$ D. equals $+\ 1$ D. as the new sphere, and we change the sign from $-$ to $+$, and the axis from 180° to 90°, as shown on the blackboard.

Now we will take the reading formula for this gentleman, and transpose according to this rule.

$$\frac{\begin{array}{ll}+\ 2\ \text{D. S.}\ \supset\ -\ 2\ \text{D. cyl. axis } 90° \\ -\ 2\ \text{D.} \qquad \text{change sign and axis}\end{array}}{0 \qquad +\ 2\ \text{D. cyl. axis } 180°}$$

In this case the algebraic addition of $+\,2$ D. and $-\,2$ D. equals nothing and leaves us without a sphere. Change the sign from $-$ to $+$ and the axis from 90° to 180°, and we have this simple cylinder as the result of the transposition.

If we wished to change this simple cylinder to a sphero-cylinder, we can do it by the same rule of transposition as follows:

$$
\begin{array}{l}
\text{o S.} \ \bigcirc\ +\ 2\ \text{D. cyl. axis 180°} \\
+\ 2\ \text{D.} \qquad\quad \text{change sign and axis} \\
\hline
+\ 2\ \text{D. S.}\ \bigcirc\ -\ 2\ \text{D. cyl. axis 90°}
\end{array}
$$

For the new sphere we add sphere and cylinder together, which is simply adding nothing to 2 D., and the result is $+\,2$ D. for the sphere, while we change the sign and axis from $+$ to $-$, and from 180° to 90.°

I have now given you the two chief rules of transposition, viz., the transposition of a cross-cylinder into a sphero-cylinder, and the transposition of a shero-cylinder into another form of sphero-cylinder. And this case is interesting as illustrating both these rules and affording a practical application of them.

If this gentleman had been willing to wear bifocals, we would simply have ordered $+\,1$ D. S. segments added to his distance lenses.

In leaving this subject it will be interesting to state that a sphere may be considered as composed of two cylinders of the same sign and power with axes at right angles; and conversely any two cylinders of the same sign and power with axes at right angles, are equal to a sphere of the same sign and power, as for example:

$$
\begin{array}{l}
+\ 1\ \text{D. S.}\ =\ +\ 1\ \text{D. cyl. axis 90°}\ \bigcirc\ +\ 1\ \text{D. cyl. axis 180°} \\
+\ 1\ \text{D. cyl. axis 90°}\ \bigcirc\ +\ 1\ \text{D. cyl. axis 180°}\ =\ +\ 1\ \text{D. S.}
\end{array}
$$

In other words we consider the refractive power of the two chief meridians, and this shows the power of the whole lens.

Adjustment of Spectacles

[CLINIC No. 33]

Mr. C. H. R., aged fifty years, complains of difficulty in reading and close work. In answer to our question he says he has never worn glasses.

On directing his attention to the test card hanging across the room, he names every letter on the No. 20 line, showing his acuteness of vision to be $\frac{20}{20}$ or normal. This, of course, excludes myopia, but there may be hypermetropia or a slight degree of hypermetropic astimatism.

In order to determine the probability of the presence of the latter, we turn to our ophthalmometer, which shows only the usual amount of overlapping in the vertical meridian.

We now make the test-case examination and place a pair of $+ 1$ D. lenses before the eyes, with which vision still remains $\frac{20}{20}$. We increase these lenses to $+ 1.50$ D., with which vision is slightly blurred and the letters on the same line are named with difficulty. We place a pair of $-.25$ D. before these $+ 1.50$ D. and now the letters are clear and the line is easily read. This proves the presence of hypermetropia to the amount of 1.25 D.

We hand patient the reading card and find that he is unable to read any of the smaller type, and for the larger type, which he is able to see, his near point is twelve inches.

This man, then, is a hypermetrope, and as he has reached the age of fifty, he is, in addition, a presbyope. The wonder is that he should have reached this age without having felt the imperative need of glasses. No doubt he did, but he ignored the demands of nature, but inevitably at the expense of a great strain on his eyes.

This is a matter of common observation : it seems natural for people to postpone the wearing of glasses as long as possible, and especially do presbyopes. But it is useless and foolish to fight against nature ; rather, it is a part of wisdom to yield easily and save the eyes the great burden of strain otherwise imposed upon them.

Our patient wants glasses only for reading, which we will figure out according to the presbyopic rule as follows :

Subtract the lens representing the receded near point from the lens representing the desired near point, and the result will be lens required to correct the presbyopia and restore the receded near point to the normal distance.

We usually select the smaller size type to determine the near point, and where such type cannot be read without glasses, we supply a known convex lens which will permit of such type to be seen and a near point to be measured. But in this case we will content ourselves with the near point for the larger type, as this will doubtless show a lens as strong as it would be wise to give to a person who has not heretofore worn glasses.

We have, then, a near point of twelve inches—will some gentleman tell me what lens represents this distance?

"Three and a quarter diopters."

Yes, that is right; a lens of thirteen inches focal distance, which is equivalent to 3.25 D.

Will some other gentleman tell us what lens represents the desired near point?

"Five diopters."

Correct; we must have a near point of eight inches, which is equivalent to 5 D.

Now, then, I will write the problem and the result on the blackboard·

$$\begin{array}{r} 5.00 \text{ D.} \\ 3.25 \text{ D.} \\ \hline + 1.75 \text{ D.} \end{array}$$

The glass representing the receded near point (3.25 D.) is subtracted from the glass representing the desired near point (5 D.).

Theoretically, then, + 1.75 D. lenses would be right to afford good vision of the medium-size type at the proper reading distance. We place this number in the trial frame and hand it to the patient. He tells us these lenses make reading very clear and distinct, and that he can even read the very small print at the top of card.

In answer to our question, he tells us he wants rimless spectacles, and I will now proceed to take the necessary measurements for the same, using this opportunity to give you a few hints on the subject of

MEASUREMENTS FOR SPECTACLES

Your work is not finished when you have completed the examination of the eyes of your patient and determined the amount and character of the error of refraction present. The correcting lenses must be set in suitable frames or mountings, which will not only be durable, but of attractive appearance and neatly fitted to the face, so that the glasses may prove satisfactory in every way not only to the wearer himself, but to his family and friends.

If you do not possess a good practical knowledge of the proper placing of glasses before the eyes, the benefits that are naturally to be expected from a careful measurement and an accurate correction of errors of refraction will be wanting.

Of late years many advances and improvements have been made in the manufacturing of comfortable and handsome contrivances for holding glasses before the eyes, and these are all the more necessary in order to meet the increasing use of prismatic and cylindrical lenses.

Some of your patients will have decided views of their own as to what they want in the way of a frame or mounting; sometimes they are right, sometimes wrong. In the latter case you should have such a knowledge of frames and mountings as to be able to clearly explain to your patient the falsity of his ideas. Many other patients will come to you with no fixed ideas, and then you are free to select for them down to every detail just what seems to be best suited to their peculiarities.

Sometimes when you have spent a good deal of time in working out an intricate case of refractive error, and when perhaps there are several patients waiting for you, there may be a temptation to slight this important branch of your art. You may feel that you have done good work in estimating the refraction, and that everything else will take care of itself. You may take it for granted that the patient must see properly with the lenses which you have spent so much time in formulating when once they are placed before his eyes, forgetting perhaps for the time that no matter how accurate a correction the lenses represent, they lose part of their effect or have an improper effect, if they do not occupy the proper position before the eyes. Hence such lenses may not only fail to do good, but may even do actual harm. Such being the case, the importance of the proper adjustment of the lenses becomes a matter of great importance.

In these days when the optometrist is striving to place himself on a plane above the ordinary seller of spectacles, you should perfect yourself as well in the mechanical branch of the work as in the scientific. One supplements the other and either is incomplete without the other. This calls for careful measurements of the face, and when the glasses are ready for delivery ample time should be given to their satisfactory adjustment, and the patient impressed with the importance of keeping them so, and requested to return as often as necessary for that purpose.

TWO STYLES OF SPECTACLES

As you already know, there are two general divisions of the mechanical construction that holds the lenses in place, viz., frames and mountings. The word frames applies to that form in which the edges of the lens are beveled so as to enter a grooved wire which surrounds it and holds it in position. This is the strongest form of spectacle, and should always be prescribed for children and workmen whose occupation renders glasses liable to breakage.

The word mountings is used for the supporting parts of "rimless" or "frameless" or "skeleton" glasses. These are preferred, especially by the ladies, for their beauty and inconspicuousness. The edges of the lenses should be finished dull, as otherwise if polished they reflect the light unpleasantly. The one great disadvantage of frameless spectacles is their liability to breakage, on account of the weakening of the glass at the points where they are drilled for the mountings, and they lack the support and strength which is afforded by a rim around the lenses.

PUPILLARY DISTANCE

The normal position for which glasses should be fitted is when the eyes are directed straight forwards, the visual axes correspond in position with the center of the lenses, the optical and geometrical centers coinciding. Glasses are sometimes purposely decentered to gain their prismatic effect, but ordinarily we fit them so that the center of the pupil lies directly behind the optical center of the lens.

If the object looked at was always in the same position and at the same distance, this would be a simple matter; but as the pupils are 4 to 6 mm. closer when looking at a near object than in distant vision, we must compromise somewhat between that position of the eyes in which the glasses will be most used and the position in which they will be less used.

When glasses are to be used for near work only, they may be decentered in 2 or 3 mm. on each side from the normal position for distance, or, what amounts to the same thing, the pupillary distance of reading spectacles should be less than if intended for distant vision ; 4 mm. for a working distance of fifteen inches, 6 mm. narrower for a distance of ten inches.

Fortunately, it is only in strong lenses that any considerable amount of prismatic effect is developed by slight decentering, and here the greatest care must be observed. But in the large number of cases where low-power lenses are worn, any possible prismatic effect is not marked.

The most important measurement, then, is the pupillary distance ; that is, the distance from the center of the pupil of one eye to the center of the other. This can be measured by a ruler held

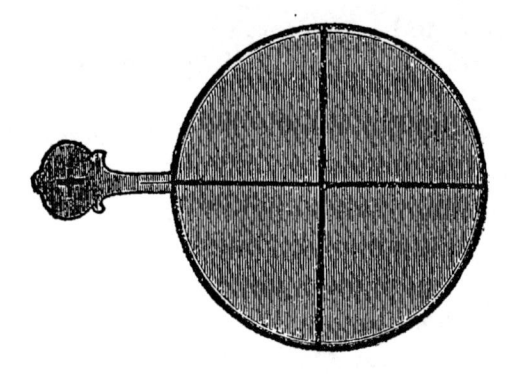

before the eyes, the patient being requested to look at some distant object, while you take your position in front and to the side of patient. The edge of the rule should be marked in millimeters or sixteenths of an inch, and the zero mark placed at the inner border of one iris, while the thumb is moved along the edge until it comes to the outer border of the other iris, where the pupillary distance can be read off. This gives the pupillary distance more accurately than trying to measure from the center of one pupil to the center of the other, because the line between the iris and the sclerotic affords a better basis for measurement. The average distance is about 60 mm. or 2⅜ inches.

Inasmuch as you are necessarily somewhat close to the patient in taking this measurement, the apparent distance will be a little less than the actual distance. Therefore, it is well to add about 2 mm. to the reading of the rule, in order to obtain the true interpupillary distance.

Or have the room darkened and the patient look at a small light across the room, when a small brilliant light will be seen in the center of each cornea. These corneal reflections mark the visual axes of the eyes, and the distance between them can be easily measured.

In addition to these methods, I am accustomed to use a trial frame and a plano lens before each eye, marked with cross lines, as shown in the illustration on the preceding page.

The thumb-screw is then rotated, thus approximating or separating the lenses, until the point of crossing of the lines is directly in front of the center of each pupil. The frame is then removed and the pupillary distance is read off the scale or measured by the rule.

BRIDGE DIMENSIONS

The next point to be considered is the *height of bridge*, which is the distance from a line passing through the centers of the lenses to the lower edge of the bridge.

If a rule be held horizontally before the patient's eyes with its edge resting on the bridge of the nose and at the natural position for the spectacle bridge, the height of this edge of the rule above the pupils will show at a glance about how high the top of the bridge should be.

I will now proceed to take these measurements on the patient before us, so that you may see exactly how it is done. I will ask the gentleman to look out of the window at extreme distance. I stand to one side so as not to obstruct his view, and place the zero of the rule at inner edge of the iris of right eye, and run my thumb-nail along the edge until it reaches the outer edge of the iris of the other eye, where I read 63 mm., which is about two and one-half inches. This is the pupillary distance for distant vision, but as this gentleman desires to use his glasses for reading only, we will deduct 4 mm., making the p. d., which we will write in the prescription, 59 mm.

I will now place the rule once again on the bridge of the nose and compare its edge with an imaginary line joining the centers of the two pupils. In this case, as you see, the edge of the rule is above this line, and, taking another rule, we will measure it (vertically, of course) and find it to be about a quarter of an inch, or 6 mm.

The inclination of the bridge is the next point to be determined. Once again we bring our rule into use, placing it in the same position as heretofore. The patient is requested to wink, and we notice what relation the tips of the lashes bear to the edge of the rule. If they sweep over this edge, then the bridge must be back of the plane of the glasses, or, in other words, the inclination of the bridge must be "in," so as throw the lenses out. If the lashes barely escape the edge of the rule, then the bridge at its top and the lenses must be on the same plane, and the inclination of the bridge is "o." If the tips of the lashes are back of the edge of he rule, then we note how much closer they may be brought to the edge of the rule without touching. This indicates that the top of the bridge of the spectacles must be in front of the plane of the lenses, or, in other words, that the inclination of the bridge must be "out."

We take this measurement in the case before us and find that the tips of the lashes are about 2 mm. back of the edge of the rule, and we will, therefore, order the inclination of the bridge 2 mm. out. You will understand when the nose is prominent, the inclination of bridge must be out; while in the case of a flat nose, the inclination of the bridge must be in.

The width of the base of the bridge is the next measurement. This is the width of the nose where the bridge at its widest opening will touch it, and is usually back of the plane of the lenses. This can be obtained by a pair of blunt compasses, the points of which are placed where the shanks of the bridge would rest. In this case we find it to be 18 mm.

We have now taken the four principal measurements for spectacle fitting, viz

Pupillary distance,
Height of bridge,
Inclination of bridge,
Width at base.

As a matter of fact, the best way to obtain or to verify these dimensions is by means of a set of "fitting frames," which are made to correspond with standard sizes of bridges, and numbered and lettered accordingly. It has been recommended that this letter and number then be given in order to save time in taking measurements and writing prescriptions, but I would advise you to compare the four measurements you desire with those afforded by the fitting frame you have selected, and write each one of the measurements separately, thus affording you an opportunity to make any alteration in either one of them that may be necessary.

We try the fitting frame marked N 2, and placing it on patient's face, it seems to be a fairly good fit. Now let us examine each measurement in detail.

The p. d. of this frame is 60 mm., but for reasons already mentioned we wish to order 59 mm. The height of this bridge is ¼ inch, but this does not allow the lenses to sit quite low enough for reading, and I think we had better increase this distance and order 8 mm. for height of bridge. The inclination of this bridge is $\frac{1}{16}''$ out; this holds the lenses a very slight distance from the tips of the lashes, so we will order 2 mm. out as we originally found it. The base of this bridge is $\frac{12}{16}$, which is equal to about 19 mm. This appears to be a very comfortable fit, and so will order 19 mm. instead of 18 mm. as per our original measurement.

It has taken me some time to describe these procedures and make these measurements, but with a set of fitting frames and a rule and a little experience, they can be taken satisfactorily in a minute or two.

Fitting of Eyeglasses

[CLINIC No. 34]

Miss Kate C., aged 31 years. She tells us that she has been employed at a dressmaker's, but that her eyes troubled her so much she was compelled to give up sewing and seek a position at domestic service. She has never worn glasses.

We find acuteness of vision in right eye to be $\frac{20}{60}$ and left eye $\frac{20}{80}$.

The ophthalmometer shows an excess of curvature in the 80th meridian of 2 D. for the right eye, and of 3 D. in the 70th meridian for the left eye. This indicates the probable presence of an astigmatism with the rule.

We hand her the reading card in order to determine the position of the near point and the amount of amplitude of accommodation. She is unable to read the smallest type, but she can see the second size, marked .75 D., with which her near point is eight inches, equivalent to an amplitude of accommodation of 5 D.

Perhaps the thought may strike some of you that this is all right, as you recall the standard of near point and accommodation which has been fixed for presbyopia. But I hasten to say that such reasoning does not apply in this case because this patient is but little over 30 years of age, and therefore cannot be presbyopic.

What inference can be drawn and information gained from this near point of eight inches? It shows a refractive power less than normal and raises at once the presumption of hypermetropia.

Will some gentleman tell us what is the normal near point and amplitude of accommodation at this age?

"Five and one-half inches, equaling 7 D. of amplitude of accommodation."

Yes, that is correct: we should have 7 D. of accommodation, but we have only 5 D. There is therefore a deficiency of 2 D. in the accommodative power, most probably due to a hypermetropic condition of refraction. Therefore it is fair to assume that the astigmatism revealed by the ophthalmometer is of the hypermetropic variety, or that there is some hypermetropia combined with it.

We will now proceed with our test case examination, commencing with the right eye as usual. Following our routine we try a + .50 D. sphere. This is neither accepted nor rejected, it does not make the letters any better or any worse. We now try a + .50 D. cylinder, placing the axis at 80° as indicated by the ophthalmometer. This is at once accepted as improving vision, thus verifying the presence of astigmatism at least to this amount.

With this cylinder in place we hold in front of it first a + .50 D. sphere and then a + .50 D. cylinder with axis in same position, at the same time asking the patient which she prefers. She unhesitatingly chooses the latter, thus making the cylindrical value in front of her eye equal to a + 1 D. We remove the + .50 D. cylinder and replace it with a + 1 D. cyl. axis 80° which improves vision to $\frac{20}{30}$ almost.

We repeat the same procedure, viz., placing a + .50 D. sphere and then a + .50 D. cylinder in front of the + 1 D. cylinder in the trial frame, and ask the patient which affords the greatest improvement in vision. She prefers the cylinder, which makes the cylindrical value + 1.50 D. We place a + 1.50 D. cylinder in the trial frame, which affords a vision of $\frac{20}{20}$. We once again repeat our former procedure of trying a + .50 D. sphere and a + .50 D. cylinder, but this time both are rejected. We try a + .25 D. sphere and a + .25 D. cylinder, but they also are rejected. We are therefore justified in assuming that this is a case of simple hypermetropic astigmatism with the rule amounting to 1.50 D. We rotate the cylinder 10° or 15° to the left, and then 10° or 15° to the right, in each case producing a marked impairment of vision, thus proving that 80° is the proper position for the axis of the cylinder, and verifying the reading of the ophthalmometer.

We now take up the examination of the left eye, going through exactly the same procedures, and the result we find to be + 2 D. cylinder axis 70°, with which vision equals $\frac{20}{30}$. This is the best we can do for the left eye. You will frequently meet with cases where it is impossible to raise the vision in both eyes to normal, and very many times it is the left eye that falls below the standard.

In order to make our examination complete, we will make use of the ophthalmoscope, retinoscope and muscle tests, but we will pass these over for the present, in order to devote some attention to eyeglass fitting.

THE FITTING OF EYEGLASSES

This lady, like most of her class, objects to wearing spectacles. We tell her that they will afford her more satisfaction and that her nose is scarcely suited for eyeglasses, in spite of which she insists on the latter; and as it always pays to please the ladies, we will make an effort in that direction.

The measurements for eyeglasses are in some particulars the same as for spectacles, and yet there is quite a difference. There are no temples or bridge, but in place of these we have the guards and spring, which are made in many varieties and sizes; in fact the market is flooded with various guards, springs, and devices for holding the glasses in place, and it would be as impossible for me to mention and describe them, as it would be for you to keep on hand samples of all of them.

The pupillary distance will be measured in exactly the same way as for spectacles, but in the case of eyeglasses it must be gotten by bending the shank of the guards, or by increasing the length of the posts or the size of the eyeglasses.

SIZES OF EYES OR LENSES

The various shapes and sizes are designated as follows: 4, 3, 2, 1, 0 and 00, number 4 being the smallest and 00 the largest. Specially large sizes are 000, 0000 and jumbo. The No. 4 is 34 mm. wide, and there is a gradual increase of 1 mm. until we reach the 00 eye which is 40 mm. wide. The jumbo or largest size is 47 mm. wide.

The size of the lenses should be in keeping with the size of the face and the inter-pupillary distance. 00 eye is the usual size for a male adult, 0 eye for a female adult, and 2 eye for a child. An 00 eye is usually suitable for a pupillary distance of $2\frac{1}{2}$ inches, an 0 eye for a p. d. of $2\frac{1}{4}$ inches, 1 eye for a p. d. of two inches, and a 2 eye for a p. d. of less than two inches. The tendency in these days is to order lenses of larger sizes than formerly, which I think is to the advantage of the patient's sight.

By a little measurement we are able to determine whether any certain size lens is suitable for any particular patient. To do this we must subtract the width of the bridge (or in eyeglasses the distance between the inner edges of the two lenses) from the pupillary distance, and the result will be the space at our disposal for the inner halves of the two lenses, which is just equivalent to the total width of one lens. If the amount of space is not sufficient, the size of the eye must be lessened, while if there is space to spare the eyes can be made larger if it is thought desirable.

For instance, suppose you had chosen an ∞ eye as suitable for the face, and you found the width of the bridge to be 20 mm., and the p. d. 60 mm. This leaves 40 mm. for the inner halves of the two lenses, which is just right because the size of an ∞ eye is 40 mm. in frameless. Whereas, if the bridge took up more than 20 mm. of the 60 mm. p. d., then the increase in the width of the

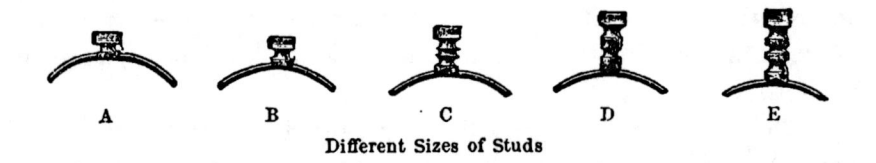

A   B   C   D   E

Different Sizes of Studs

bridge must be deducted from the size of the lenses. In other words, the size of the lenses must be made to correspond to the available space: we can lessen the size of the lenses, but we cannot lessen the width of the nose as shown by the bridge.

In the case of eyeglasses we can increase the p. d. by increasing the size of the lenses, as for instance a No. 2 eye is 36 mm. wide while an ∞ eye is 40 mm. wide, and therefore we gain 4 mm. in p. d. by using the latter.

In addition we can control the p. d. in eyeglasses by the different lengths of studs. There are five standard lengths, marked A, B, C, D and E, the A being the shortest and the E the longest. The A stud is about 3 mm. long and the E stud about 9 mm. long, so that by using the latter we can gain 6 mm. on each side; in other words, we can increase the p. d. 12 mm. by using the E studs, or any intermediate gain in p. d. by using one of the other lengths of studs.

THE SPRING

The spring comes in different shapes and sizes: oval or round, semi-oval and flat, and varying in length from 1¾ inches to 2¾ inches. The width of the nose where the guards are to fit should be taken, in order to get some indication as to the size of spring: if the nose is narrow a shorter spring will suffice than if the nose is large and broad at its base.

The choice between a round or flat spring is largely a matter of taste, although in my opinion a round spring looks best on a full face and a flat spring best on a small or narrow face.

The nearness of the lenses to the eyes can be regulated by the guards, by the spring and by the studs.

The guards. If the bearing surface of the guards is made to slant obliquely outwards from front to back, the lenses will be allowed to approach closer to the eyes. If the guards are bent so that the bearing surface is straighter, the lenses will be thrown a little farther from the eyes.

The position of the lenses can also be regulated by the length of the arm of the guards. If the arm is long the lenses are held farther away : if the arm is short, the lenses are brought closer to the eyes. The same thing can be accomplished by so-called wide-angle and close-angle guards.

Out of the many kinds of guards on the market, we should select the simplest and the one that will permit of whatever bending is necessary to adapt the bearing surfaces to the slope of the nose. The guard that meets the requirements in the greatest number of cases is the No. 1 shell guard. In the hands of an optometrist who is experienced in its adjustment, this guard can be satisfactorily fitted to nine-tenths of the noses you meet. Then there is No. 3, which is a close-angle, short-arm guard, and brings

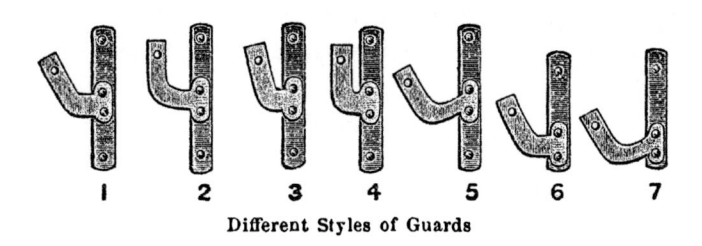

Different Styles of Guards

the lenses closer to the eyes. Also No. 5, which is a wide-angle, long-armed guard, and throws the lenses farther away. As soon as you try a pair of No. 1 guards on your patient, you can see at a glance whether lenses are at right distance from eyes or whether they should be closer or farther, in which case the indications for Nos. 3 and 5 will present themselves to you. Nos. 6 and 7 are used when the lenses set too high, and when it is desired to throw them down as for reading.

The guards are lined with shell or cork and such have heretofore been almost exclusively used. But the tendency at the present day is to prefer the all-metal guards as more cleanly and more sanitary. These unlined guards can be more easily adjusted and bent, as there is no danger of breaking the shell or defacing the cork.

The Spring. The "Grecian" spring, which curves outwards, allows the lenses to come closer to the eyes than the straight spring. The weight of the spring depends somewhat upon the strength of the lenses: if the power is low and the lenses light, a reduced or light weight spring will answer; whereas, if the lenses are strong and thick and heavy, a more substantial spring may be needed.

The Studs. When the Grecian spring will not allow the lenses to come close enough to the eyes, then we can make use of so-called "inset studs," which can be procured of several lengths to bring lenses $\frac{1}{16}''$ or $\frac{1}{8}''$ closer as may be desired. Or if the eyes are projecting and it becomes necessary to hold the lenses farther away, the same studs can be used by reversing their direction, making them "outset studs."

The height of the lenses can be regulated by the guards and by the position in which the holes are drilled for the mountings.

The Guards. This depends upon the position at which the arm is attached to the guard. The usual place is in the middle of the guard where it holds the lenses in proper position for the majority of persons. If this is too low, the attachment of the arm to the upper part of the guard will raise the lenses: if too high, the attachment of the arm to the lower part of guard (as in Nos. 6 and 7) will lower them.

Position of holes. Ordinarily the hole is drilled on a plane with the center of the lenses, which answers for the majority of persons. If this allows patient to see under the lenses, or if the glasses are desired for reading only, the lenses may be ordered "drilled above center," $\frac{1}{16}''$ or $\frac{1}{8}''$ as may be necessary, thus lowering the glasses this much.

But the most practical way to take the measurements for eyeglass mountings is to try them on the nose, and for this purpose it is necessary to have a comprehensive set of samples. These should be fitted with plano lenses and neatly made, so as to make a good impression on the patient on whom they are tried. The eyes should be of various shapes and sizes, and it is a good idea to have them marked with a vertical and a horizontal line crossing each other at the center of the lens. These sample fitting eyeglasses should show the different lengths of studs, offset and inset studs, and also the various styles of springs and the guards that are most commonly used.

With such a set of sample fitting eyeglasses, embracing the most important styles of guards, studs, springs, and the several sizes of eyes, it is a comparatively easy matter to select the one that comes nearest to holding the lenses in the proper position and to maintain a secure hold on the nose and at the same time feel easy and comfortable.

In order that the eyeglasses may keep their proper position without slipping, it is important that the guards bear uniformly on the nose along their whole surface, with just a little slope to make the anterior edge of the guards fit slightly closer. Also that the tops of the guards may be approximated a trifle so as to bring the pressure above the bridge of the nose in the fleshy portion and not on the bone, which makes it more comfortable and more apt to stay on. At the same time care should be taken to see that the lenses are properly centered both vertically and horizontally.

It is really a fact that scarcely any pair of spectacles, much less eyeglasses, will fit perfectly without a certain amount of adjustment. No matter how carefully you take your measurements, how correctly you write the dimensions on the prescription or how accurately the manufacturer fills your order, yet when you get the glasses and try them on the patient's face, they may fail to be a perfect fit.

Has there been any error on your part or the manufacturers? Probably not. The final adjustment must be made on the face that is to wear the glasses. For instance, if a pair of eyeglasses were ordered with the guards 14 mm. apart at the top and 20 mm. apart at the bottom, and the slant of the nose was a little different on one side from the other, the glasses would be tilted up on the side that was the fullest, and the set of the glasses would be wrong even though the nose dimensions as given were correct.

Or in the case of spectacles: if one ear was a little higher than the other or one side of the head a little fuller than the other, the glasses will not be straight, or one lens will stand farther from the face than the other.

One other practical point: the glasses may be adjusted correctly at the first sitting, and yet get out of adjustment later on as they settle to their permanent resting place on the face. Therefore, it is always well to suggest to the patient that he return in a few weeks for a refitting or sooner if necessary, as otherwise he may condemn you for not having made a good job of the fitting.

There are two ways of adjusting glasses, with the fingers and with the pliers. It is wonderful what some persons can do to frames and mountings with their fingers, but for good work we prefer the pliers. There is always danger in using them, especially for frameless work, even if you are skilled. The best way is to have the measurements in your prescription so accurate that your need for pliers will be reduced to a minimum. This applies when you send your orders to the manufacturer, but if you wish to fill from your own stock you may have to depend a great deal on the pliers.

It is unsatisfactory to work on steel with pliers as it is likely to snap. In gold filled goods the use of pliers is apt to crack or break the thin outer coating of gold. Gold is the best material to work on with pliers, as the metal, while tough, is also pliable, and gives way under the pressure of the tools.

In the use of pliers the motion should be slow and gentle, with the avoidance of any sudden twists, and producing the effect desired by several applications of the pliers.

We now turn to our patient and carry out the principles I have just given you. We find the pupillary distance to be $2\frac{3}{8}$ inches, and I will look among our fitting eyeglasses, selecting this one, which looks as if it might fit. It has a pupillary distance of $2\frac{3}{8}$ inches, with No. 1 shell guards and No. 11 spring. I place it on the patient's nose, and you see the left glass tilts upwards. I take my pliers and bend out the bottom of the guard thus lowering the lens, which now appears straight. I will also bend the arms of the guards a trifle so as to correspond to the slope of the nose, and replacing the glasses you see they are a perfect fit.

Before parting, one more word about the pliers. The man back in the shop who is using these tools all the time, can do a great deal more of altering of frames and mountings without spoiling their appearance than you can out in your office, and therefore you should not use them any more than actually necessary. Of course you must do some adjusting, but the measurements given in the prescription should be so nearly accurate that everything possible in the mechanical line should be done by the man who is accustomed to do this kind of work.

Spectacles and Eyeglasses

[CLINIC No. 35]

Mrs. S. L. R., aged 60, complains of some headache, a tired feeling in the eyes after reading, and recently eyes have been somewhat inflamed.

We find the acuteness of vision in each eye to be $\frac{20}{60}$.

The ophthalmometer shows an excess in the vertical meridian of 1 D.

With the retinoscope a + 2 D. neutralizes the motion in the vertical meridian, and a + 2.50 D. in the horizontal meridian.

The ophthalmoscopic examination is negative—the refracting media are all clear but a slight dullness in the appearance of the fundus.

The test case examination develops the following combination, + 1.50 D. S. ○ + .50 D. cyl. axis 90°, with which vision equals $\frac{20}{20}$.

For reading there is required the addition of + 2.50 D. S., making the reading formula, + 4 D. S. ○ + .50 D. cyl. axis 90°.

This then is a case of compound hypermetropic astigmatism complicated with presbyopia, calling for two pairs of glasses.

She decides she will have spectacles for constant wear, and eyeglasses for reading, thus giving us an opportunity to put to practical use the principles of spectacle and eyeglass fitting which I have demonstrated at the last two clinics, and to throw out a few more hints along the same lines.

When you meet a person who would like to have eyeglasses instead of spectacles, but whose nose is so flat and shallow that eyeglasses will not hold, it will probably save you a good deal of future trouble if you do not attempt to fit them but insist on spectacles. Such persons often have false hopes kindled in them by the advertisements of opticians who, in exploiting the claims of certain styles of guards, make the extravagant statement that they can be fitted to any nose.

When you are fitting a person from your sample stock of eyeglasses, it is not always well to show the various styles to your patient and leave the selection to him, as he may arbitrarily select a style which cannot be fitted to his nose. But rather decide for yourself what kind seems best adapted to his nose and proceed to fit as you think best. You will be apt to get more satisfactory results and save time that would have to be spent in argument and explanation as to why certain styles are not suitable for that particular case, and besides the patient may have more confidence in your ability if his opinion is not asked.

I will now proceed to take our measurements, first for the spectacles. This lady's face and features are as you see small and narrow. I take my rule and laying its edge on the bridge of the nose, according to the directions I have given you at a previous clinic, I find the p. d. to be $2\frac{1}{8}$ inches. The edge of the rule is about a quarter of an inch above the imaginary line connecting the centers of the pupils. The nose being somewhat sharp and slightly prominent, the edge of the rule is a little distance from the tips of the lashes, so that the inclination of the bridge will be $\frac{1}{16}''$ out. The width of the base is $\frac{5}{8}''$.

Generally a frame can be selected from stock that will afford the desired measurements, but not always. One frame may have the correct p. d. but not the right height of bridge : while another that has the proper height of bridge, may be found too wide or too narrow. In such cases the dimensions which are correct can be taken from various frames, in this way making up the prescription that possesses the required data to be sent to the wholesale house.

Occasionally you may meet with a face on which some of the dimensions cannot be obtained from any of your fitting frames, or there may be a difference between the two sides of the face. Then it becomes necessary to take the measurements as I have already shown you.

There is no frame in our collection that affords all the measurements desired in this case. This frame which I hold in my hand is marked "M $1\frac{1}{2}$," the dimensions of which are as follows : p. d. $2\frac{1}{4}$; height $\frac{3}{16}$; inclination $\frac{1}{16}$ out ; base width $\frac{5}{8}$. Two of these dimensions are correct, the inclination of the bridge and the width of its base ; but the p. d. and height are incorrect, the one being too much and the other too little. So it is with the rest of the frames; where some of the dimensions are right, the others are wrong. However we will try this frame on the lady's face, and note each of the measurements desired.

The first glance shows that it is too wide and out of proportion to the small face, so that our former measurement of 2⅛ inches is verified. We notice that the lenses are too high and patient looks through their lower edges, so that height of bridge should be a little more than in this frame, increasing from $\frac{3}{16}''$ to $\frac{1}{4}''$. The lenses just escape the tips of the lashes, so that we know the inclination of the bridge is right. The bridge fits the nose neatly, being neither too tight nor too loose, so that we assume our measurement of width of base is correct.

In regard to the bridge you should always see that when the bridge rests at its proper place that the shanks do not press uncomfortably into the inner canthi. If the shape of the bridge is such as to produce this effect, then a bridge with short shanks should be ordered. In other cases long shanks may be required. The inclination of the bridge is also affected by the shanks, long shanks adding $\frac{1}{16}''$ and extra long shanks $\frac{1}{8}''$.

For instance, in this case if this frame allowed the lashes to brush slightly against the lenses: instead of changing the inclination of the bridge to throw the lenses farther from the face, I could order M 1½ bridge with long shanks, which being $\frac{1}{16}''$ longer, would hold the lenses that much farther from the eyes and thus escape the lashes. Or if the lashes brushed the lenses a good deal, I could order M 1½ with extra long shanks, which would set the lenses $\frac{1}{8}''$ farther out than the regular dimensions of the bridge.

In addition to the four dimensions which I have described to you and which I have deemed the most important, there are four others which must sometimes be taken into account.

The angle of the crest,
The angle of the lenses,
The width of temples,
The length of temples.

The angle of the crest. This is a measurement that is usually neglected, probably because we take it for granted that all noses where the bridge of the spectacles rests, show the same angle to the face. But this is not right, and it is really of importance that the slope of the nose should be observed so as to give us some idea whether the usual angle of crest will suffice or whether some change should be made. Otherwise the sharp cutting edge of the crest breaks the skin of the nose in a short time, making an ugly sore that it is impossible to heal while the glasses are worn with the angle of crest in same position. The patient also becomes irritated and is apt to have an unpleasant feeling for the man who furnished the glasses.

If the bearing surface of the bridge of the nose is more nearly vertical than normal, the lower edge of the crest will cut into the nose. Or if the slope of the nose is more nearly horizontal, the upper edge will cut into the flesh.

A neat little instrument has been devised for measuring the proper angle of the crest, or it can also be determined by a card marked with radiating lines which show the angle of the slope of nose in comparison with a plane lying horizontally.

The angle of the crest then you will understand has reference to the angle formed by the inside surface of the bridge where it rests on the nose with a horizontal line. 30° would be a small angle, 60° a large angle, while 45° would fit the average nose.

The angle of the lenses. The perpendicularity of the plane of the lenses to the visual axis is an important matter, the rule being that the lenses shall set as nearly as possible at right angles to the visual axes. The stronger the lens the more the importance of this rule is emphasized. The reason for this lies in the fact that the refractive value of the lens is changed and increased when placed obliquely to the axis of vision.

A cylinder when misplaced in this way gives the effect of a stronger cylindrical lens; a sphere not only an increased spherical effect but also a cylindrical effect, the axis being at right angles to the meridian about which the lens is rotated.

A 1 D. cylindrical lens at an obliquity of 30° gives the effect of an additional half diopter; while at an obliquity of 40°, the cylindrical effect is just doubled.

A 1 D. sphere at an obliquity of 40° gives an increased sperical effect of .16 D. with the addition of a cylindrical effect of nearly 1 D.

In order to meet the requirements of perpendicularity to the visual axis, lenses used only for distance should be vertical as shown in diagram on blackboard.

In near vision which is usually below the level of the eyes, the visual axes are directed downwards and inwards, and here the lenses must face in the same direction in order to maintain the desired perpendicularity, as shown in the diagrams on the blackboard.

When glasses are prescribed for constant wear, the lenses may be placed midway between the adjustment for distance and for near. In this case the inclination is not exactly correct for either purpose, but at the same time the obliquity to the visual axis in either case is so slight as to cause but little inconvenience, because it is only when the obliquity is marked that the increase in power and the development of a cylindrical effect in spherical lenses is noticeable. Of course a moderate degree of obliquity of the lenses can be neutralized by a slight bending of the neck of the wearer.

This question of the perpendicularity of lenses is a most important one when bifocal glasses are prescribed, and these should nearly always be ordered tilted to the position proper for near use. It has frequently been my experience that when patients return with bifocals complaining that they were not satisfactory, a re-examination would show that the lenses were correct, and on further investigation it would be found that the lenses were too nearly vertical and that a slight tilt as for reading was all that was necessary to make them comfortable and satisfactory.

When the temples are perpendicular to the plane of the glasses, the latter will face directly forward as for distant vision. When it is desired to make them face downwards as for reading, the necessary obliquity may be obtained by simply turning down the temples, care being taken that they are equally turned, as otherwise if one is turned more than the other the lens on this side will be higher on the face than on the other.

In cases also where the lashes are long and where it is difficult to get the reading glasses close enough to the face, the patient will probably see under the glasses when he looks down for reading. This difficulty can be overcome by bending the temples as just described so as to make the lenses tilt inwards at the bottom. In most cases a tilting of 10° will suffice, but this may be increased to 15° or 20° if thought necessary.

The width of temples. The temple width refers not to the distance from joint to joint across the frame, but to the distance between points on the temples one inch back from the front, the reason for this being that the temples come in contact with the sides of the head about one inch back from the plane of the lenses, and that is the "width" we desire to measure.

This measurement becomes necessary where the patient has a broad face with perhaps no or only a slight increase in pupillary distance. In such a case if the p. d. only is specified in the prescription, on account of the disproportion in the face, the temple wires will cut into the sides of the head. If the face is narrow and the temples stand out from the head, the glasses are likely to be unsteady when walking or in exercise, and especially if they are heavy. In order to give the proper support to the lenses, the temples should simply touch the skin without any pressure. An occasional patient will complain if there is the slightest contact between the temple and skin, when it may become necessary to bend the temples out a trifle even at the expense of losing the support which the glasses would otherwise have.

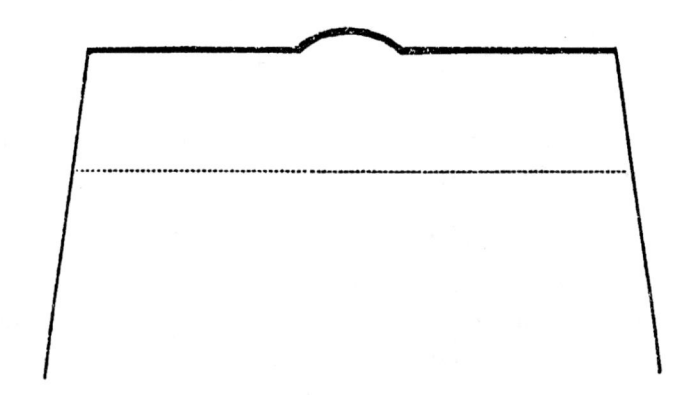

A glance at the diagram I have made on the blackboard will perhaps serve to make my meaning still clearer.

The length of temples. When this measurement is mentioned, some persons think it refers to the distance from the temple joint to the point where it begins to curve down over the top of the ear, which varies from 3½ to 4 inches. But I think it would save confusion to give the length of the entire temple.

The only reason why the length should be given from joint to top of ear, is that the manufacturer may put a sharp curve at this point, as illustrated on blackboard.

But unless specified in the prescription the manufacturer will not put this kink in the temples as it is not customary to do so. In spite of this there are some authorities who claim that the proper form for hook temples is a straight line from the joint to top of ear, then a sharp curve and finally an easy curve corresponding to back of ear. Therefore, if in any case you should consider it desirable to have a temple of this shape, you had better do the bending yourself after the frame comes from the hands of the manufacturer. With patience and a little experience, the temple wire can be bent to follow every depression and elevation of the surface with which it lies in contact; this is especially desirable in any case where the auricle is deformed or irregular in shape.

The ordinary length of the entire temple is six inches and this will be found to answer in the majority of cases. If you do not specify any particular length of temple in your prescription, you will receive one of six inches. For children it will be best to order one 5½ inches long, or in small young children one of 5 inches. On the other hand if the face is full and the head large, a temple 6½ or 7 inches long is often necessary.

DEFECTS EASILY REMEDIED

When you receive the glasses and fit them on your patient, if the temples are a trifle too long, and you do not wish to lose the time to return them to be cut off, you may curl the tip of the temples back a little so that the ball on the end will not press uncomfortably into the flesh behind the ear. I do not recommend this procedure except where the excess in length is very slight, as otherwise this curl on the end of temples may be unsightly and distasteful to the patient.

If the glasses are inclined to slip down the nose, the curve of the temples should be shortened by drawing them between a penholder (or any smooth round object) and the finger, pressure being made against the *inner* side of the curve. While if the glasses are drawn too tightly toward the eyes, the curve may be lengthened by drawing the temples between the finger and penholder, the pressure in this case being made against the *outer* side of the curve.

You should be careful to see that both temples stand at the same angle as illustrated on the blackboard ·

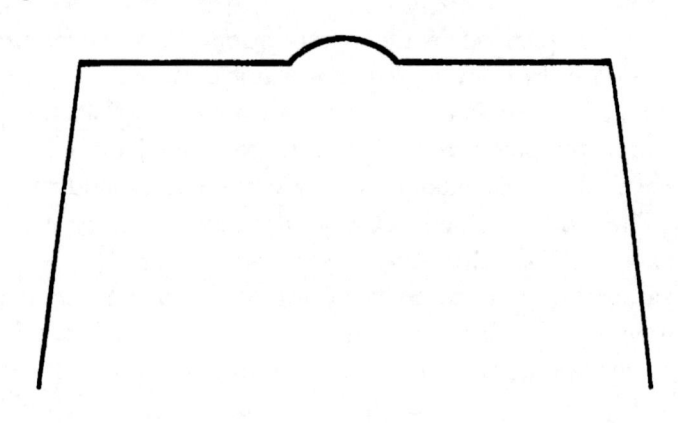

If one temple forms a larger angle, the lens on that side will be drawn closer to the face, so that the lashes or even the lids may touch the glass.

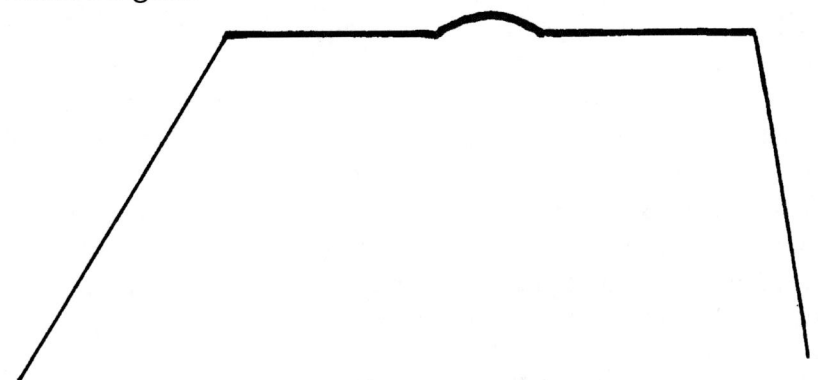

If the ears are both on the same level, each temple must make the same angle with the front, otherwise one lens may be too high and the other too low, or if one ear is higher than the other, as not infrequently happens, one temple must be raised or the other lowered to overcome the difference. If the temple is bent down near the joint, the lens on this side will be made to stand higher, while if the temple is bent up, the lens on this side will be lowered.

If you find the glasses do not set straight, you can raise one lens or lower the other as may seem best to you, by bending the temple down or up with round-jawed pliers used close up the joint. If both temples are turned downwards, the effect is not to raise both lenses but to tilt them forwards as for reading, as I have previously told you.

All I have said about length of temples refers of course to riding bows or hook temples; when straight temples are desired (which is seldom now-a-days) the length is not specified in the prescription, except perhaps to indicate whether the ordinary length will suffice or whether they should be extra long.

Inspection of Spectacles and Eyeglasses and Neutralization of Lenses

[CLINIC No. 36]

It is a matter of prudence and importance that you should carefully examine the finished product as received from the manufacturer in order to determine if your prescription has been faithfully filled. Neglect of this precaution may nullify the results of your patient and skilful work in measuring the error of refraction. Therefore it becomes absolutely necessary for you to know how to test and neutralize lenses correctly and expeditiously, as well as to correct any mal-adjustment of the frame or mounting that becomes apparent when placed on the patient's face.

INSPECTION OF SPECTACLES

In the case of frames examine the lenses to see if they are in properly. You know that a lens can move in the eye wire so that one end or the other will tilt up or down, which in the case of cylinders becomes a serious matter. Take your screwdriver and loosen the screw, turn the lens to its proper position, and then tighten the joint.

Turn the glasses sidewise to see if they are both on the same plane : if not, there is probably a twist in the bridge, which can be rectified by the use of two pliers, one for holding and the other for turning. If there is a concave surface on the lens (as in periscopic convex) see that it is next to the eye. If both surfaces are concave, see that the strongest concave is next to the eye.

Place the glasses on patient's face to see if the angle of the lenses is right ; nearly vertical for distance, or tilted for reading. See that the temples work smoothly in the joints : sometimes the loosening of the screw or the application of a drop of oil becomes necessary. Note the angle of the crest of the bridge and see that it corresponds to the slope of the nose. See that the width of the base of the bridge is right so that it fits the nose neatly. If it is too tight or too loose, the half-round pliers can be used to widen or narrow it, which if much will necessitate bending of the shanks of the bridge in order to keep the lenses in alignment.

Note if the pupillary distance is as ordered, and see if the temples fit comfortably against the side of the head: the wires must not cut into the flesh, neither must they stand away from the skin. Either of these faults can be remedied by bending temples close to joints, but care must be taken to see that both temples are at the same angle, else the glasses will not be equally distant from the eye.

If the ears are sensitive and if in spite of loosening the curve of the temple as much as allowable, complaint is made of the pressure on the cartilage of the ear, in which case you may slip a piece of fine rubber tubbing over the wires to prevent the cutting, or use a special device now on the market for this purpose.

ADJUSTING EYEGLASSES

As I have told you before the guards of eyeglasses should conform to the slope of the sides of the nose, with a little extra tightness at the tops to prevent tilting forwards. Then attention should be given to the spring to see if it is too tight or not tight enough, or if the tension is just about right. You can tighten the spring with half-round pliers and weaken it with flat-nosed pliers. If the guards fit properly, it is not necessary that the spring should be tight; in fact, we want to avoid all unnecessary pressure as otherwise your patient will soon have a sore nose. Another point is to see that the spring is of the proper length, and this is something that is often overlooked. When you alter the tension of the spring, you will find it necessary also to readjust the guards.

If the lenses of eyeglasses droop, they can be raised by bending the spring just where it enters the stud closer to the lenses, and at the same time bending the lower parts of the guards inwards. If the lenses tilt up, the spring should be bent away from the lenses and the lower parts of the guards bent outwards.

If one of the lenses sets farther from the eye than the other, you should look to see if the fault does not lie in the bearing surfaces of the guards, when probably a little slant made in the faulty one will bring the lens into proper position.

DETERMINING THE STRENGTH OF LENSES

In the inspection of a lens the first point to be determined is whether it is convex or concave. If the lens is strong the convex or concave curves are so marked that as soon as we look at the lens we can see its character. But in the weaker lenses that are in common use, we are accustomed to determine this point by the apparent motion caused by the lens.

I take the lens in my hand, holding it eight or ten inches from my eye and looking through it at the letters on the distant test card. I then move the lens from side to side, and this will cause an apparent motion in the letters looked at. If the motion is in the same direction as the lens is moved, it is concave; if the motion is opposite, convex. By this means it is easy to detect as low a power as .12 D. and any one can use it without any special skill or preparation.

Instead of moving the lens from side to side, it may be moved closer to and farther from the eye. If the object looked at magnifies as the lens is pushed away from the eye, it is convex; if it grows smaller, concave. There is one precaution that should be observed in both of these methods, and that is in the case of convex lenses, which must be held within their focal distance, or the results will be reversed.

Having determined in this way if the lens is convex or concave, the next step is to ascertain if it is simple or compound, and whether cylinder or prism is present. Holding the lens in your hand at a distance from your eye as before, look through it at some object presenting a straight line, as the window sash or a picture frame.

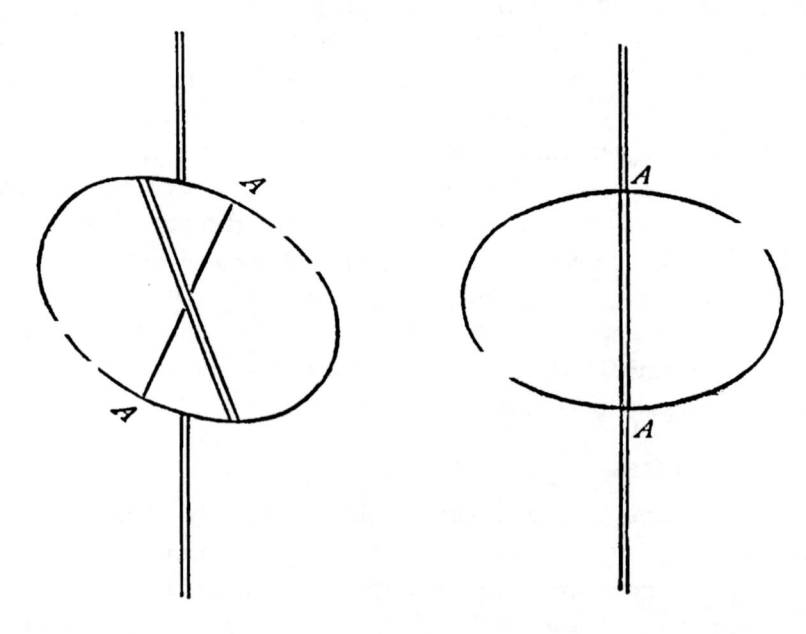

Rotate the lens around the visual line acting as a longitudinal axis, and note the effect on the straight line which you have selected for observation.

PROVING PRESENCE OF A CYLINDER

If that portion of the line seen through the lens appears to move, or in other words if there is a break in the continuity of the line seen through the lens as compared with that above and below it, then a cylinder is proven to be present. If this oblique displacement is in a direction contrary to the motion of the lens, the cylinder is convex: if in the same direction as the motion, concave. The drawing on the blackboard indicates the action of a convex cylinder. To locate the position of the axis of the cylinder, it should be slowly rotated until the line seen above, below and through the lens is continuous, as illustrated on the blackboard. The line would also be continuous in the meridian at right angles to the axis; in other words, the line is continuous only in the chief meridians of the lens, viz., the meridians of least and greatest refraction. In the case of a weak cylinder (.12 D. or .25 D.) the oblique motion produced is slight and you must watch closely to detect it. In a cylinder of higher power (1 D. and over) the effect is very pronounced. I would advise you all to get a few cylindrical lenses of various strengths and spend some little time in practicing with them until you become proficient in the detection of a cylinder.

In looking through a lens at a straight line, there is one word of caution I wish to give you, and that is not to confound lateral displacement with oblique displacement. In looking through a simple sphere at any place except its optical center, the straight line will be broken, but the broken lines will be parallel, and they can be made to coincide not by rotating the lens but by moving it from side to side. But in the case of a cylinder the displacement seen through the lens is oblique and is caused not by moving the lens from side to side, but by rotating it.

TO DETERMINE SPHERICAL LENSES

The majority of lenses are spheres, either simple or compound, that is, at least one surface is spherical in curvature.

Sometimes we meet with plano cylinders, in which lenses there is no power in the meridian of the axis, all the refractive value being in the meridian at right angles. In testing to determine if a lens is a simple sphere, we may use the straight line again, rotating the lens around its optical center and taking care to keep the latter in alignment with the straight line : if no "twisting" action is developed, there can be no cylinder present.

DETECTING PRESENCE OF A PRISM

To determine a prism. The presence of a prism in a lens is usually disclosed by simple inspection and noticing a difference in thickness in the two opposite edges of the lens.

In the absence of a prism the edges of the lens should be of the same thickness at opposite points. Of course, if the prism is of low degree, it may escape detection on a casual examination; but we have a very simple method by which we may determine its presence. Hold the lens in the fingers as previously described and look at the same straight line directly through the optical center. If the line is continuous above, below and through the center, the absence of a prism is proven. But if the line is broken, a prism is present, the displacement of that portion of the line seen through the lens being in the direction of the apex.

Proving the optical center. Before handing the finished glasses to your patient, each lens should be carefully examined with reference to the position of its optical center and also the distance between the two, as otherwise an error in these important particulars may be overlooked. In the cheap glasses sold by peddlers and in the five and ten-cent stores, proper centering is probably the exception, such glasses being known as second-class or even third-class.

In first-quality lenses, unless otherwise ordered, we assume that the lenses are properly centered and that the optical center and the geometrical center coincide, as otherwise our measurements for pupillary distance, however carefully made, are vitiated, and an undesired prismatic effect introduced into the lenses. Therefore it becomes necessary for you to be able to locate the optical center of a lens, which can be easily done as follows :

Place a rectangular card on the table or tack it on the wall ; hold the lens some distance from it and from your eye. The edges of the card seen through and outside of the card, will appear continuous only when the corner of the card is exactly at the optical center of the lens. This I have illustrated in the diagram on the blackboard, No. 1.

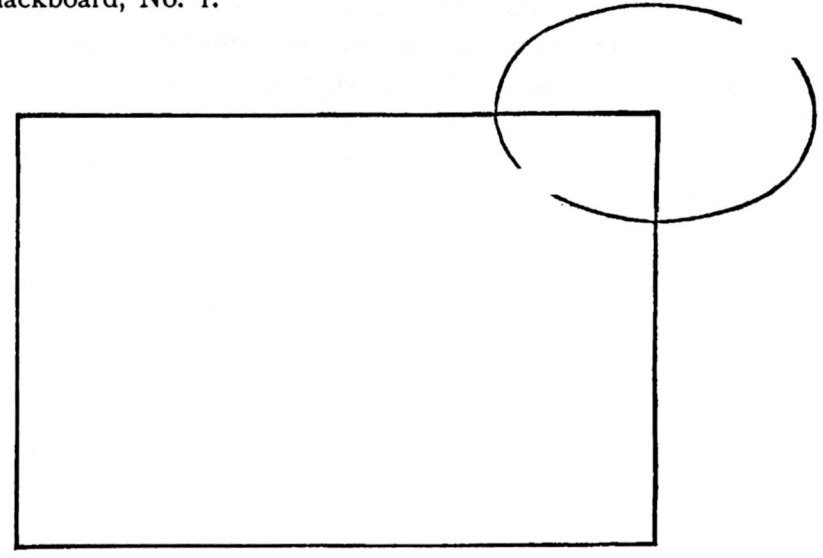

No. 1

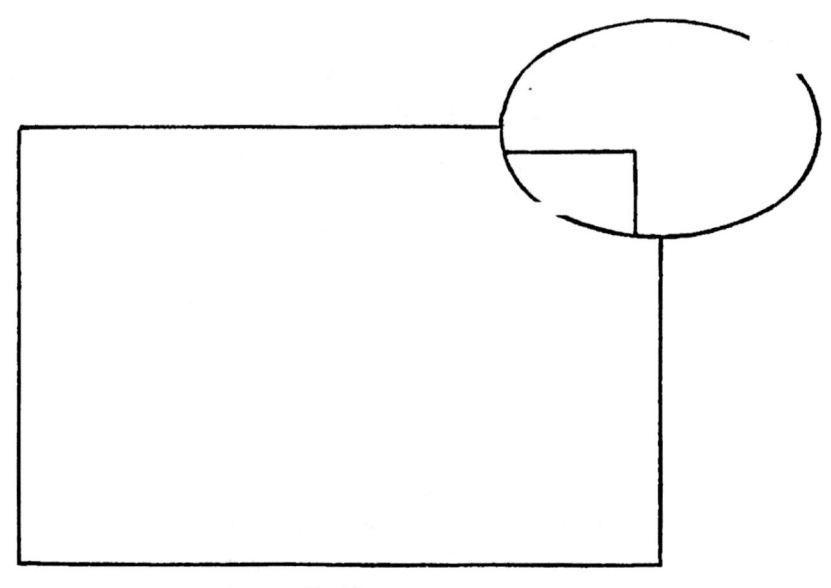

In diagram No. 2 the lens is improperly held in relation to the card, while in No. 3 the lens being correctly held, the optical center is shown to be displaced downwards and sideways. If this method is used with care, the results are satisfactorily accurate.

No. 2

Neutralization of spherical lenses. To neutralize is to nullify or make of no effect. As I told you a few moments ago when we look through a convex or a concave lens in motion, a certain effect is produced on the object looked at, causing it to move against or with. When we neutralize the lens we destroy or stop all such effect or motion. Neutralization is the most common method of

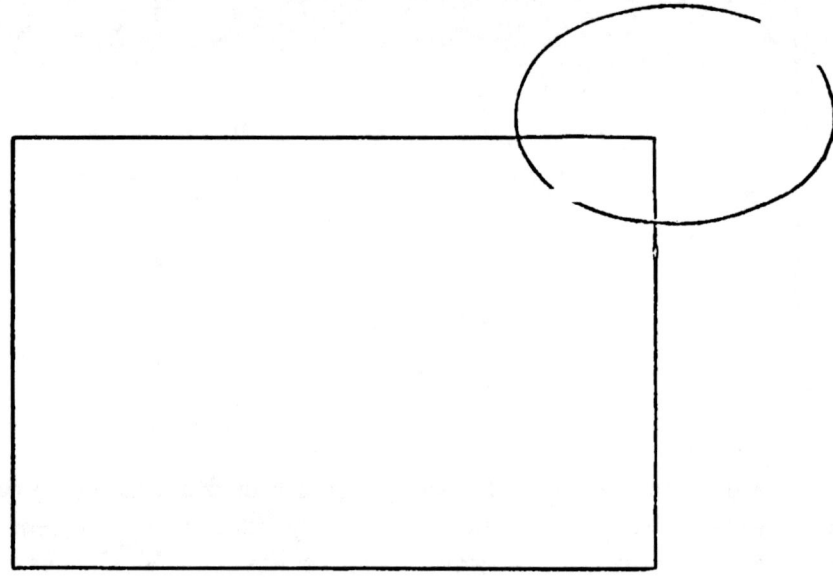

No. 3

measuring the strength of lenses. In taking a lens in your hand for this purpose, you first determine whether it is convex or concave by the method I have already described to you, and at the same time you gain some idea of the strength of the lens, because the more rapid the movement of the object looked at, the stronger the lens.

If the lens in your hand is convex, you take from the trial case a concave of the estimated strength, and place them in apposition center to center, and make a trial of the combination as you previously made of the single lens, and carefully note the apparent movement.

1. If the motion is still opposite, then the concave lens you have chosen is too weak, and another and stronger must be tried.

2. If on the contrary the motion is now in the same direction as the two lenses are moved, then your concave lens is too strong and a weaker one must be tried.

3. If no motion is apparent the neutralization is perfect, and you are looking through what corresponds to a plano lens, because the power of the convex lens has been destroyed by the concave and the value of the first lens can be read off the handle of the second lens, always taking the precaution to change its sign.

Therefore, to determine the strength of a spherical lens, you combine it with successive lenses of opposite sign from the trial case until one is found that checks all motion. In lenses of high power we watch for the neutralization at or near the center of the lens, as motion and distortion may still be noticeable near the periphery.

Neutralization of cylindrical lenses. You have already determined that the lens is cylindrical according to the method previously explained to you and also its nature, whether convex or concave. In moving the lens you have discovered the meridian in which there is no motion, which indicates the position of the axis, and the neutralizing cylinder taken from the trial case must be placed with its axis in exactly the same position. If the lens under test is a convex cylinder, then you must try successive concave cylinders until one is found that stops motion in all directions. If a concave cylinder 1.50 D. with its axis vertical is required, then the lens under examination is $+$ 1.50 D. cyl. axis 90°.

A sphero-cylindrical lens. This is a compound lens composed of a sphere and a cylinder, and its neutralization is a matter of some difficulty, especially for beginners. With such a lens there is motion in all directions, but you will soon find that there is one meridian in which motion is least rapid. Take from the trial case a sphere of the proper sign and strength to neutralize the movement in this meridian. You will find there is still motion in the meridian at right angles; in other words, by the use of the sphere to neutralize one meridian, you have now in your hand what amounts to a plane cylinder, which you proceed to neutralize according to the method I have just described to you. The neutralizing lenses you have now in your hand are a sphere and a cylinder, which represent the value of the lens you are testing. For instance, if your neutralizing lenses are $+$ 1.50 D. sphere and $+$ 1 D. cylinder with axis held horizontally, then the compound lens you are testing is $-$ 1.50 D. S. $\bigcirc$ $-$ 1 D. cyl. axis 180°. In a case like this where two lenses are required for neutralization, you will at first find some difficulty in handling all three of the lenses at once and keeping the axis in the proper position.

Neutralizing Prisms. The strength of a prism may be expressed in two ways: by its refracting angle or by its power to bend a ray of light from its course. The latter system has advantages over the former, but this is a point which we have not the time to discuss at present.

A prism may be neutralized by another prism taken from the trial case and placed in apposition, the base of one over the apex of the other. A straight line viewed through the prism is broken, that part seen through the lens being deflected towards the apex. The neutralizing prism with base over apex would bring the line back and if of the proper strength make it continuous.

Neutralization affords a most satisfactory method of determining the refractive value of a lens, whether simple or compound, but it has its disadvantages. Sometimes you may get a perfect neutralization by a sphere and cylinder both of which are concave, thus showing convex values, and yet by the feel and shape of the lens you know that one of its surfaces is concave. Let me illustrate by writing two formulæ on the blackboard:

$+ 1.50$ D. S. $\bigcirc + 1$ D. cyl. axis $90°$
$+ 2.50$ D. S. $\bigcirc - 1$ D. cyl. axis $180°$

By following the method I have described to you of neutralizing the weakest meridian first and then the other meridian by a cylinder, you will get

$- 1.50$ D. S. $\bigcirc - 1$ D. cyl. axis $90°$

and you assume that the lens is the one first written ($+ \bigcirc +$), but as you examine the surfaces of the lens you can see that the inside surface is concave cylindrical. Now the fact of the matter is that this is all in transposition. The two lenses whose formulæ are written on the blackboard have the same refractive value, one being transposable into the other, and by means of neutralizing lenses you can determine only the refractive value of the combination, but not the curvature of each surface.

The Lens Measure. This leads me to mention the lens measure, a little instrument which I would advise you all to purchase, as affording the quickest and most convenient method of determining the strength and composition of lenses, although at the same time I would insist that you should first be a master of the art of neutralizing lenses from the trial case.

You are all familiar with the appearance of a lens measure, and you probably all know that of the three projecting pins the two outside ones are stationary, while the central one being movable shows the amount of curvature which is indicated by the hand on the dial.

I have frequently handed a lens and the measure to a young student and asked him to tell me the strength of the lens. He presses it against one surface, the hand moves around to 2.25 on the convex side, and he answers "plus 2.25." "That is not correct," I say. He tries it again, the hand stops at the same place, and he looks at me in bewilderment when I shake my head no. Now, what is the trouble, or wherein does the error of the student lie?

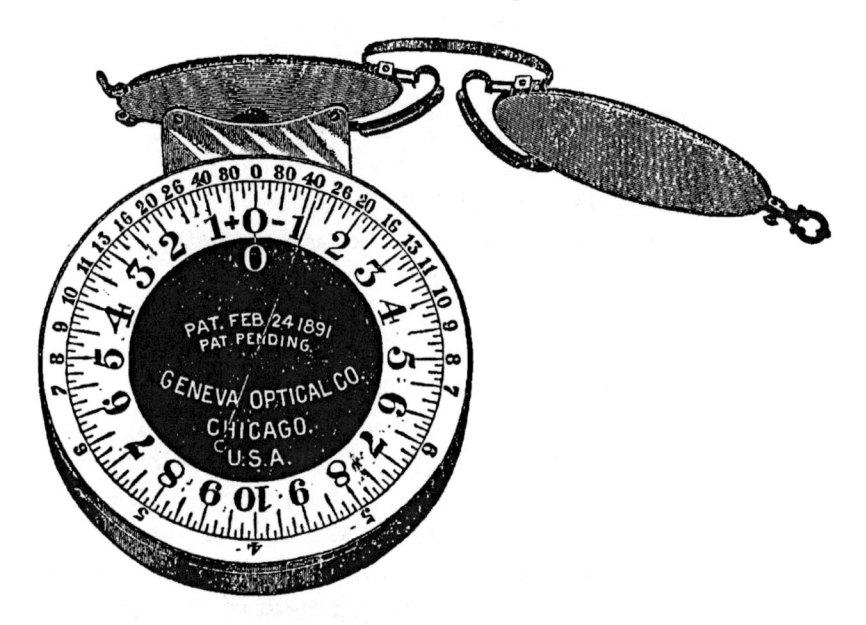

Let me say to you and emphasize it, that every lens has two surfaces and both of them must be taken into account. The trouble with our friend was that he measured one surface only.

Both surfaces of the lens must be measured separately and then their values combined by means of algebraic addition: Sometimes both surfaces have the same curvature, as in bi-convex or bi-concave, sometimes one surface is convex and the other concave, as in periscopic, and sometimes one surface spherical and the other cylindrical as in sphero-cylinders.

Now, then, taking the lens which our friend tried to measure, on pressing the measure against the other surface we find it to be

— 1.25 D., which as you know is the standard for the concave surface of periscopic convex lenses. Now then by algebraic addition, we have

$$\begin{array}{r} + 2.25 \text{ D.} \\ - 1.25 \text{ D.} \\ \hline + 1 \text{ D.} \end{array}$$

This lens then is a periscopic lens whose value is + 1 D.

The lenses in the trial case are double, that is, the same curvature on both sides. I will take one and measure it, finding + 1.25 D. on both sides.

$$\begin{array}{r} + 1.25 \text{ D.} \\ + 1.25 \text{ D.} \\ \hline + 2.50 \text{ D.} \end{array}$$

By algebraic addition we find the value of this lens is + 2.50 D.

Keeping the measure steadily pressed against the lens, I rotate it through all the meridians of the lens at the same time watching the dial : if the pointer remains stationary, the curvature of the surface is spherical, its value being indicated by the figures at which the pointer stops. Both surfaces are tried in the same way.

If, however, a rotation of the measure causes the hand to move, you will know that you have to do with a cylindrical surface. You turn the measure to the point where the hand points to zero, and the three points are then standing on the axis of the cylinder. Turning the measure again you see the power begin to increase until at right angles to the axis, the full strength of the cylinder is shown. If you are measuring a plano cylinder, the opposite surface will be plane and the hand will point to zero in all meridians. If the lens is a sphero-cylinder, you must measure both surface to discover its refractive value. If the lens measure shows + 1 D. in all meridians on one surface, and + 1 D. in the horizontal meridian and zero in the vertical meridian on the other surface, the lens is + 1 D. S. ◯ + 1 D. cyl. axis 90°.

INDEX

A

Abduction, 26, 107
Accommodation, spasm of, 191
Adduction, 26, 107
Adjusting of glasses, 230
    of spectacles, 215
Albuminuric retinitis, 177
    acuteness of vision in, 177
    characteristic features of, 181
    impairment of vision in, 177
    in pregnancy, 179
    optic disk in, 177
    significance of, 179
    use of ophthalmometer in, 177
    use of pin-hole disk in, 177
Amblyopia, 66
    acuteness of vision in, 67
    hysterical, 29
    preventable, 66
    use of prisms in, 67
    use of trial case in, 67
Amblyopia, toxic, 35
    ophthalmoscopic examination in, 36
Aniridia, 145
Anisometropia, case of, 78
    acuteness of vision in, 84
    causes of, 81
    considerations in correcting, 83
    examination of patient, 7
    impaired vision in, 78
    ophthalmometric examination in, 79
    ophthalmoscopic examination in, 79
    optic disk in, 79
    use of Maddox rod in, 80
    use of test lenses in, 79
    varieties of, 82
Astigmatism against the rule, 134
    compound hypermetropic, 135
    compound myopic, 136
    cornea in, 136
    definition, 135
    mixed, 136
    ophthalmometric examination in, 138
    simple hypermetropic, 135
    simple myopic, 136
    test-case examination in, 138
Astigmatism, case of, illustrating value of ophthalmometer, 115
Astigmatism, trial-case test in, 119
    compound hypermetropic with presbyopia, 46
Astigmatism, hypermetropic, simulating myopic, 20
    acuteness of vision in, 20
    fogging test in, 22
    Maddox rod in, 21
    muscular equlibrium in, 21
    testing for, 20
    use of prisms in, 22
Astigmatism, lenticular, 141
    abnormal, 142
    cases of, 144
    cornea in, 142
    exceptional cases of, 148
    seat of, 142
Astigmatism, mixed, 71
    cause of, 207
    cycloplegic in, 75
    definition of, 73
Astigmatism, mixed
    examination of patient, 72
    fogging test in, 74
    stenopaic slit in, 76
    subjective tests of, 75
    test letters in, 75
Astigmatism of the lens, dynamic, 145
Astigmatism, simple hypermetropic, 9
    Maddox-rod test in, 12
    trial-case test in, 10
    with normal vision, 9
    with the rule, 11
Astigmatism, symmetric, with the rule, 118
Astigmatism, routine of examination in, 121
Astigmatism with the rule, 128
    compound hypermetropic, 130
    compound myopic, 131
    cornea in, 129
    correction of, 131
    direct, 129
    examination of patient in, 131
    explanation of, 129
    hypermetropic, 129
    indirect, 129
    mixed, with the rule, 131
    myopic, 129
    simple hypermetropic, 130
    simple myopic, 130
    use of ophthalmometer in, 129
Asthenopia, 23
    exophoric, 204
Atropine, 26

B

Bifocals, 236
Blind spot, 174
Bridge, the, 220, 233
    height of, 220
    inclination of, 220
    measurement of, 220
    width of base of, 221
Bright's disease, the eye in, 178

C

Cataract, 41
    definition of, 41
    incipient, 42
    pupil in, 41
    refractive error diagnosed as, 41
    revealed by ophthalmoscope, 41
Choked disk, 180
Choroidal atrophy, 148
Choroidal ring, 174
Ciliary muscle, 11, 33, 193
    contraction of, 124
    in myopia, 61
    in spasm of accommodation, 193
    relaxation of, 124
Circles of diffusion, 35
Color vision in toxic amblyopia, 38
Concave lenses, 116, 152
    decentration of, 152
    use of, 116
Conical cornea, 155
    cause of, 161
    iridectomy in, 163
    myotic in, 163

Conical cornea
    testing for, 155
    use of ophthalmometer in, 156
    use of pin-hole test in, 156
    use of trial lenses in, 156
Conjunctiva, hyperaemia of, 194
Conjunctivitis, 196
Contact glasses, 162
Conus, the, 150
Convergence, deficiency of, 200
Convex lenses, 115
    effects of strong, 124
    use of in examination, 115
Corneal curvature, normal, 30
Crescent, angle of, 233
Cross cylinder, 73
Crystalline lens, removal of, 153
Cylinder, proving presence of, 243
Cycloplegia, simulated, 74
Cycloplegic, use of, 123

D

Dark room, 172
Decentration, 152
    rule for, 153
Diplopia, artificial, 80
    artificial vertical, 202
    crossed, 203
    homonymous, 12, 22, 185
    heteronymous, 80
Divergence, power of, 15
Duction test, 17

E

Esophoria, 12, 17, 127, 184
    accommodation in, 184
    acuteness of vision in, 184
    cases of, 184
    headache in, 185
    measurement of, 185
    method of testing for, 184
    of low degree, 16
    prisms in, 189
    symptoms of, 188
    treatment of, 188
Exophoria, 107, 127, 199
    acuteness of vision in, 199
    case of, 199
    causes of, 201
    diagnosis of, 202
    effect of lenses on, 27
    muscular insufficiency in, 200
    prismatic lens in, 204
    prism exercise in, 205
    symptoms of, 200
    treatment of, 204
    use of Maddox rod in, 200
    use of ophthalmometer in, 199
    use of retinoscope, 199
Eye, cataphoric, 123
    hypermetropic, 32
    hyperphoric, 113
    mental picture of the, 137
    myopic, 61
Eyeglass fitting, 223
Eyeglass, inspection of, 240
    measurement of, 228
    mountings, 228
Eyeglasses, simple fitting, 229
Eyes, different sizes of, 225

F

Facultative hypermetropia, 103, 122
Fitting eyeglasses, 223
Fitting frames, 221
Fogging method in hypermetropia, 32
    in myopia, 126

Fovea centralis, 174
Fusion faculty, the, 88

G

Guards, 227
    different styles of, 227

H

Headache and eyestrain, 30, 97
    asthenopic, 99
    causes of, 98
    glasses as a cure for, 98
    in connection with myopia and exophoria, 97
Heterophoria, detection of, 21
    latent, 16
    mixed, 14
Hypermetropia, 103
    absolute, 105
    acuteness of vision in, 103
    detecting presence of, 103
    facultative, 105
    fogging method in, 106
    illustrating fogging method, 122
    latent, 105, 123
    Moddox rod in, 127
    manifest, 105
    measuring amount of, 103
    muscular equilibrum in, 32
    muscle testing in, 106
    objective examination in, 104
    prescribing glasses for, 107, 127
    simulating myopia, 63
    symptoms of, 103
    testing for, 104
    typical cases of, 30
    use of ophthalmometer in, 122
    use of retinoscope in, 33
Hypermetrope, sight of the, 31
Hyperphoria, 109
    acuteness of vision in, 109
    aetiology of, 112
    examination for, 109
    fogging system in, 110
    headaches in, 111
    latent, 112
    Maddox rod in, 111
    manifest, 17
    muscular equilibrium in, 111
    prescribing prisms for, 113
    retinoscope test in, 110
    treatment of, 113
Hysteria, 25, 191
Hysteria and eyestrain, 27
Hysteria, eye manifestations of, 29

K

Keratoconus, 155
    cause of, 161
    iridectomy in, 163
    myotic in, 163
    testing for, 155
    use of ophthalmometer in, 156
    use of pin-hole test in, 156
    use of trial case in, 156

L

Lachrymation, 194
Lenticular astigmatism, 141
Lens, sphero-cylindrical, 73, 247
Lenses, angle of, 234
    cylindrical, neutralization of, 247
    determining kind of, 242
    determining strength of, 241
    inspection of, 241

Lenses
    neutralization of, 240
    prismatic, 19
    size of, 225
    spherical, neutralization of, 246
    toric, 159
    transposition of, 208
Leucoma, 41

M

Macula lutea, 174
Maddox-rod test in presbyopia, 57
    in anisometropia, 80
    in astigmatism, 12, 21
    in esophoria, 200
    in hypermetropia, 32
    in hyperphoria, 109
Maddox multiple rod, 16
Meniscus, 162
Monocular vision, 66
    acuteness of vision in, 55
    fogging test in, 68
    impaired vision in, 66
    pin-hole test in, 65
    retina in, 66
    retinoscopic test in, 68
    tenotomy in, 66
Mydriatics, use of, 172
Myopia, 58
    definition of, 62
    false, 63
    in children, 64
    optic disk in, 59
    pin-hole test in, 60
    prescribing glasses for, 61
    subjective test for, 59
    use of ophthalmoscope in, 59
    use of retinoscope in, 59
    use of trial case in, 59
    visual acuity in, 58
Myopia, accommodative, 197
    apparent, 197
    artificial, 124
    simulative, 194
Myopia, high, 148
    accommodation in, 149
    cause of, 148
    extraction of lens in, 154
    heredity as cause of, 150
    investigating cause of, 150
Myopic crescent, 148, 150

N

Near point, the, 20
Neutralization of lenses, 246
    cylindrical, 247
    prisms, 248
    spherical, 246
Night blindness, 175
Nystagmus, 175

O

Ocular gymnastics, 206
Ophthalmometer, 15, 30, 117
    how to use, 117
    importance of, 120
    in albuminuric retinitis, 177
    in anisometropia, 78
    in astigmatism, 115, 129, 207
    in esophoria, 199
    in hypermetropia, 122
    in keratoconus, 156
    in pigmentary retinitis, 170
    in strabismus, 89, 92
Ophthalmoscope, 170
    evolution of, 170
    in cataract, 42

Ophthalmoscope
    in myopia, 59
    in pigmentary retinitis, 170
    in presbyopia, 57
    in toxic amblyopia, 35
    invention of, 171
    limitations of, 44
    Loring's, 172
    use of, 170
Optical centering, 244
Optical disk, 173
    color of, 173
    examination of, 173
    shape of, 174
Optic nerve, 244

P

Papillitis, 180
Perimeter, 37
    in strabismus, 95
    presence of, 39
    use of, 95
Photophobia, 194
Physiologic cup, 174
Pigmentary retinitis, 170
    use of ophthalmometer in, 170
    use of ophthalmoscope in, 171
    use of optic disk in, 174
Pin-hole test, 35
    in albuminuric retinitis, 177
    in keratoconus, 156
    in strabismus, 92
    in toxic amblyopia, 36
Pliers, 230
    use of in adjusting, 230
    use of in gold filled goods, 230
Presbyopia, 52
    acuteness of vision in, 53
    amplitude of accommodation in, 56
    glasses to prescribe in, 54
    illustrated case of, 53
    muscular equilibrium in, 57
    testing for, 53
    use of ophthalmoscope in, 57
    use of retinoscope in, 57
Primary position, 117
Prism exercises, 27
    effect of, 28
Prisms, 18, 22, 244.
    detecting presence of, 244
    neutralization of, 248
Pupillary distance, 218
    measurement of, 219

R

Radiating lines, 11, 75
Recti muscles, 112
Retina of a squinting eye, 86
Retinal hemorrhage, 179
Retinoscope, 25, 33
    handling mirror of, 167
    in exophoria, 199
    in hypermetropia, 33
    in hyperphoria, 110
    in myopia, 59
    in presbyopia, 57
    in strabismus, 89, 165
    use of, 165
Retinoscopy, 163
    principles of, 163
    value of, 163

S

Scleral ring, 174
Scotoma, 38
Secondary position, 118

Spasm of accommodation, 191
    ciliary muscle in, 195, 197
    diagnosis of, 191, 195
    lachrymation in, 194
    photophobia in, 194
    symptoms of, 194
    treatment of, 196
Spectacle measurements, how to make, 232
Spectacles, adjustment of, 215
    frames of, 217
    frameless, 218
    inspection of, 240
    measurements for, 217
    mountings, 217
    rimless, 218
    skeleton, 218
    styles of, 218
Spectacles and eyeglasses, 231
Spring, the Grecian, 229
Springs for eyeglasses, 226
Staphyloma, posterior, 148, 150
Stenopaic slit, 76
Strabismus, convergent, 68, 85, 164
    aetiology of, 87
    binocular vision in, 85
    cure of, 87
    definition of, 85
    deviation in, 164
    Donder's investigation of, 87
    examination of patient, 89
    fusion faculty in, 88
    objective testing of,
    use of ophthalmometer in 89
    use of prisms in, 90
    use of retinoscope in, 89
    with hypermetropia, 93, 164
Strabismus, divergent, 91
    acuteness of vision in, 91

Strabismus, divergent
    measuring the degree of, 95
    method of examination, 91
    treatment of, 93
    use of ophthalmometer in, 92
    use of pin-hole disk in, 92
    with myopia, 93
Strabismus, convergent, caused by hypermetropia, 163
    objective method of examination, 164
    retinoscopic test in, 165
    prescribing glasses for, 169
Strabismus, high convergent, 111
    latent, 186
    manifest, 186
Studs, outset, 228
    different sizes of, 226

T

Temples, 236
    length of, 237
    width of, 236
Transposition of lenses, 207
Trial frame, three-cell, 158

U

Urea in nephritis, 181

V

Vertical streak, 22
Vitreous humor, opacity of, 148

Y

Yellow spot, 188

PHYSIOLOGIC OPTICS

Ocular Dioptrics—Functions of the Retina—Ocular Movements and Binocular Vision

By Dr. M. Tscherning
Director of the Laboratory of Ophthalmology at the Sorbonne Paris.

AUTHORIZED TRANSLATION

By Carl Weiland, M. D.
Former Chief of Clinic in the Eye Department of the Jefferson College Hospital, Philadelphia, Pa.

This book is recognized in the scientific and medical world as the one complete and authoritative treatise on physiologic optics. Its distinguished author is admittedly the greatest authority on this subject, and his book embodies not only his own researches, but those of the several hundred investigators who, in the past hundred years, made the eye their specialty and life study.

Tscherning has sifted the gold of all optical research from the dross, and his book, as now published in English, with many additions, is the most valuable mine of reliable optical knowledge within reach of ophthalmologists. It contains 380 pages and 212 illustrations, and its reference list comprises the entire galaxy of scientists who have made the century famous in the world of optics.

The chapters on Ophthalmometry, Ophthalmoscopy, Accommodation, Astigmatism, Aberration and Entoptic Phenomena, etc.—in fact, the entire book contains so much that is new, practical and necessary that no refractionist can afford to be without it.

Bound in Cloth.   380 Pages, 212 Illustrations.

Price $2.50 (10s. 5d.)

Published by
THE KEYSTONE PUBLISHING CO.
809-811-813 NORTH 19TH STREET, PHILADELPHIA, U. S. A.

TESTS AND STUDIES
OF THE
OCULAR MUSCLES

By Ernest E. Maddox, M. D., F. R. C. S., Ed.

Ophthalmic Surgeon to the Royal Victoria Hospital, Bournemouth, England; formerly Syme Surgical Fellow, Edinburgh University.

This book is universally recognized as the standard treatise on the muscles of the eye, their functions, anomalies, insufficiencies, tests and optical treatment.

All opticians recognize that the subdivision of refractive work that is most troublesome is muscular anomalies. Even those who have mastered all the other intricacies of visual correction will often find their skill frustrated and their efforts nullified if they have not thoroughly mastered the ocular muscles.

The eye specialist can thoroughly equip himself in this fundamental essential by studying the work of Dr. Maddox who is known in the world of medicine as the greatest investigator and authority on the subject of eye muscles.

The present volume is the second edition of the work, specially revised and enlarged by the author. It is copiously illustrated and the comprehensive index greatly facilitates reference.

Bound in silk cloth—261 pages—110 illustrations.

Sent postpaid on receipt of price $1.50 (6s. 3d.)

Published by
THE KEYSTONE PUBLISHING CO.
809-811-813 North 19th Street, Philadelphia, U. S. A.

THE
PRINCIPLES of REFRACTION
in the Human Eye Based on the Laws of Conjugate Foci

By Swan M. Burnett, M. D., Ph. D.

Formerly Professor of Ophthalmology and Otology in the Georgetown University Medical School; Director of the Eye and Ear Clinic, Central Dispensary and Emergency Hospital; Ophthalmologist to the Children's Hospital and to Providence Hospital, etc., Washington, D. C.

In this treatise the student is given a condensed but thorough grounding in the principles of refraction according to a method which is both easy and fundamental. The few laws governing the conjugate foci lie at the basis of whatever pertains to the relations of the object and its image.

To bring all the phenomena manifest in the refraction of the human eye consecutively under a common explanation by these simple laws is, we believe, here undertaken for the first time. The comprehension of much which has hitherto seemed difficult to the average student has thus been rendered much easier. This is especially true of the theory of Skiascopy, which is here elucidated in a manner much more simple and direct than by any method hitherto offered.

The authorship is sufficient assurance of the thoroughness of the work. Dr. Burnett was recognized as one of the greatest authorities on eye refraction, and this treatise may be described as the crystallization of his life-work in this field.

The text is elucidated by 24 original diagrams, which were executed by Chas. F. Prentice, M.E., whose pre-eminence in mathematical optics is recognized by all ophthalmologists.

Bound in Silk Cloth.

Sent postpaid to any part of the world on receipt of price, $1.00 (4s. 2d.)

Published by
THE KEYSTONE PUBLISHING CO.
809-811-813 NORTH 19TH STREET, PHILADELPHIA, U. S. A.

THE OPTICIAN'S MANUAL
VOL. I.

By C. H. Brown, M. D.

Graduate University of Pennsylvania; Professor of Optics and Refraction; formerly Physician in Philadelphia Hospital; Member of Philadelphia County, Pennsylvania State and American Medical Societies.

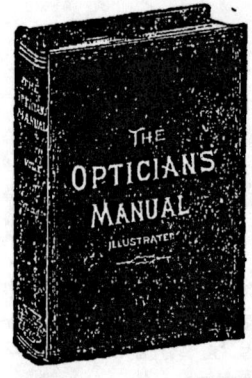

The Optician's Manual, Vol. I, was the most popular and useful work on practical refraction ever written, and has been the entire optical education of many hundred successful refractionists. The knowledge it contains was more effective in building up the optical profession than any other educational factor. It is, in fact, the foundation structure of all optical knowledge as the titles of its ten chapters show:

Chapter   I.—Introductory Remarks.
Chapter   II.—The Eye Anatomically.
Chapter   III.—The Eye Optically; or, The Physiology of Vision.
Chapter   IV.—Optics.
Chapter   V.—Lenses.
Chapter   VI.—Numbering of Lenses.
Chapter VII.—The Use and Value of Glasses.
Chapter VIII.—Outfit Required.
Chapter   IX.—Method of Examination.
Chapter   X.—Presbyopia.

The Optician's Manual, Vol. I, was the first important treatise published on eye refraction and spectacle fitting. It is the recognized standard text-book on practical refraction, being used as such in all schools of Optics. A study of it is essential to an intelligent appreciation of its companion treatise, The Optician's Manual, Vol. II, described on the opposite page. A comprehensive index adds much to its usefulness to both student and practitioner.

Bound in Cloth—422 pages—colored plates and illustrations.

Sent postpaid on receipt of $1.50 (6s. 3d.)

Published by
THE KEYSTONE PUBLISHING CO.
809-811-813 NORTH 19TH STREET, PHILADELPHIA, U. S. A.

THE OPTICIAN'S MANUAL
VOL. II.

By C. H. Brown, M. D.

Graduate University of Pennsylvania; Professor of Optics and Refraction; formerly Physician in Philadelphia Hospital; Member of Philadelphia County, Pennsylvania State and American Medical Societies.

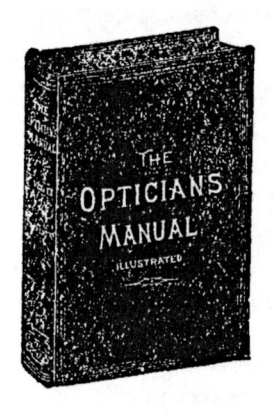

The Optician's Manual, Vol. II., is a direct continuation of The Optician's Manual, Vol. I., being a much more advanced and comprehensive treatise. It covers in minutest detail the four great subdivisions of practical eye refraction, viz:

    Myopia.
    Hypermetropia.
    Astigmatism.
    Muscular Anomalies.

It contains the most authoritative and complete researches up to date on these subjects, treated by the master hand of an eminent oculist and optical teacher. It is thoroughly practical, explicit in statement and accurate as to fact. All refractive errors and complications are clearly explained, and the methods of correction thoroughly elucidated.

This book fills the last great want in higher refractive optics, and the knowledge contained in it marks the standard of professionalism.

Bound in Cloth—408 pages—with illustrations.

Sent postpaid on receipt of $1.50 (6s. 3d.)

Published by
THE KEYSTONE PUBLISHING CO.
809-811-813 NORTH 19TH STREET, PHILADELPHIA, U. S. A.

OPHTHALMIC LENSES

Dioptric Formulæ for Combined Cylindrical Lenses, The Prism-Dioptry and Other Original Papers

By Charles F. Prentice, M.E.

A new and revised edition of all the original papers of this noted author, combined in one volume. In this revised form, with the addition of recent research, these standard papers are of increased value. Combined in one volume, they are the greatest compilation on the subject of lenses extant.

This book of over 200 pages contains the following papers:

Ophthalmic Lenses.
Dioptric Formulæ for Combined Cylindrical Lenses.
The Prism-Dioptry.
A Metric System of Numbering and Measuring Prisms.
　　The Relation of the Prism-Dioptry to the Meter Angle.
　　The Relation of the Prism-Dioptry to the Lens-Dioptry.
The Perfected Prismometer.
The Prismometric Scale.
On the Practical Execution of Ophthalmic Prescriptions Involving Prisms.
A Problem in Cemented Bi-Focal Lenses, Solved by the Prism-Dioptry.
Why Strong Contra-Generic Lenses of Equal Power Fail to Neutralize Each Other.
The Advantages of the Sphero-Toric Lens.
The Iris, as Diaphragm and Photostat.
The Typoscope.
The Correction of Depleted Dynamic Refraction (Presbyopia).

PRESS NOTICES ON THE ORIGINAL EDITION:

Ophthalmic Lenses.

"The work stands alone, in its present form, a compendium of the various laws of physics relative to this subject that are so difficult of access in scattered treatises."—*New England Medical Gazette.*

"It is the most complete and best illustrated book on this special subject ever published."—*Horological Review*, New York.

"Of all the simple treatises on the properties of lenses that we have seen, this is incomparably the best. . . . The teacher of the average medical student will hail this little work as a great boon."—*Archives of Ophthalmology, edited by H. Knapp, M.D.*

Dioptric Formulæ for Combined Cylindrical Lenses.

"This little brochure solves the problem of combined cylinders in all its aspects, and in a manner simple enough for the comprehension of the average student of ophthalmology. The author is to be congratulated upon the success that has crowned his labors, for nowhere is there to be found so simple and yet so complete an explanation as is contained in these pages."—*Archives of Ophthalmology, edited by H. Knapp, M.D.*

"This exhaustive work of Mr. Prentice is a solution of one of the most difficult problems in ophthalmological optics. Thanks are due to Mr. Prentice for the excellent manner in which he has elucidated a subject which has not hitherto been satisfactorily explained."—*The Ophthalmic Review*, London.

The book contains 110 Original Diagrams.　Bound in cloth.

Price $1.50 (6s. 3d.)

Published by
THE KEYSTONE PUBLISHING CO.
809-811-813 NORTH 19TH STREET, PHILADELPHIA, U. S. A.

SKIASCOPY
AND THE USE OF THE RETINOSCOPE

By Geo. A. Rogers

Formerly Professor in the Northern Illinois College of Ophthalmology and Otology, Chicago; Principal of the Chicago Post-Graduate College of Optometry; Lecturer and Specialist on Scientific Eye Refraction.

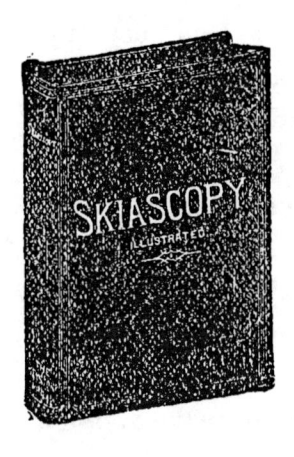

A Treatise on the Shadow Test in its Practical Application to the Work of Refraction, with an Explanation in Detail of the Optical Principles on which the Science is Based.

This work far excels all previous treatises on the subject in comprehensiveness and practical value to the refractionist. It not only explains the test, but expounds fully and explicitly the principles underlying it—not only the phenomena revealed by the test, but the why and wherefore of uch phenomena.

It contains a full description of skiascopic apparatus, including the latest and most approved instruments.

In depth of research, wealth of illustration and scientific completeness this work is unique.

Bound in cloth; contains 231 pages and 73 illustrations and colored plates.

Sent postpaid to any part of the world on receipt of $1.00 (4s. 2d.)

Published by
THE KEYSTONE PUBLISHING CO.
809-811-813 NORTH 19TH STREET, PHILADELPHIA, U. S. A.

OPTOMETRIC RECORD BOOK

A record-book, wherein to record optometric examinations, is an indispensable adjunct of an optician's outfit.

The Keystone Optometric Record-book was specially prepared for this purpose. It excels all others in being not only a record-book, but an invaluable guide in examination.

The book contains two hundred record forms with printed headings, suggesting, in the proper order, the course of examination that should be pursued to obtain most accurate results.

Each book has an index, which enables the optician to refer instantly to the case of any particular patient.

The Keystone Record-book diminishes the time and labor required for examinations, obviates possible oversights from carelessness, and assures a systematic and thorough examination of the eye, as well as furnishes a permanent record of all examinations.

Sent postpaid on receipt of $1.00 (4s. 2d.)

Published by
THE KEYSTONE PUBLISHING CO.
809–811–813 NORTH 19TH STREET, PHILADELPHIA, U. S. A.

14 DAY USE
RETURN TO DESK FROM WHICH BORROWED

OPTOMETRY LIBRARY

This book is due on the last date stamped below, or on the date to which renewed.

Renewed books are subject to immediate recall.

LD 21–50m-6,'59
(A2845s10)476

General Library
University of California
Berkeley